Monographs in Electrochemistry

Series Editor

Fritz Scholz, University of Greifswald, Greifswald, Germany

Surprisingly, a large number of important topics in electrochemistry are not covered by up-to-date monographs and series on the market, some topics are even not covered at all. The series "Monographs in Electrochemistry" fills this gap by publishing in-depth monographs written by experienced and distinguished electrochemists, covering both theory and applications. The focus is set on existing as well as emerging methods for researchers, engineers, and practitioners active in the many and often interdisciplinary fields, where electrochemistry plays a key role. These fields range – among others – from analytical and environmental sciences to sensors, materials sciences and biochemical research.

More information about this series at http://www.springer.com/series/7386

Maciej Chotkowski · Andrzej Czerwiński

Electrochemistry
of Technetium

 Springer

Maciej Chotkowski
Faculty of Chemistry
University of Warsaw
Warsaw, Poland

Andrzej Czerwiński
Faculty of Chemistry
University of Warsaw
Warsaw, Poland

ISSN 1865-1836   ISSN 1865-1844   (electronic)
Monographs in Electrochemistry
ISBN 978-3-030-62865-9   ISBN 978-3-030-62863-5   (eBook)
https://doi.org/10.1007/978-3-030-62863-5

This Springer imprint is published by the registered company Springer Nature Switzerland AG
The registered company address is: Gewerbestrasse 11, 6330 Cham, Switzerland

Preface of the Series Editor

This monograph is about the electrochemistry of technetium; however, it also contains an overview on the electrochemistry of the elements manganese, rhenium, molybdenum and ruthenium, as they share many similarities with technetium. The last review of technetium electrochemistry in a book is 14 years old and only 8 pages long (Encyclopedia of Electrochemistry, Vol. 7a, Wiley VCH, 2006). Therefore, the authors and the editor felt that an updated and much more expanded coverage is needed. The authors, Andrzej Czerwinski and Maciej Chotkowski from Warsaw University, Poland, are very well-known specialists, who were able to survey the electrochemistry of technetium and related elements because of their own research studies. I am sure that electrochemists and nuclear chemists will highly welcome this monograph.

Greifswald, Germany   Fritz Scholz
January 2020

Contents

1   **Introduction General Information on Technetium**   1
    References   7

2   **Comprehensive Electrochemistry of Tc and Its Periodic**
    Table Neighbors   11
    References   27

3   **Technetium Coordinated by Inorganic Ligands in Aqueous**
    and Nonaqueous Solutions   31
    3.1   Acidic solutions   35
    3.2   Alkaline solutions   51
    References   63

4   **Technetium Coordinated by Organic Ligands in Aqueous**
    and Nonaqueous Solutions   69
    4.1   Nonaqueous Solutions   70
    4.2   Aqueous Solutions   78
    References   103

5   **Metallic Technetium, Corrosion, Technetium Alloys**
    and Its Behavior in Spent Nuclear Fuel   109
    References   137

6   **Determination of Trace Amounts of Tc by Electrochemical**
    Methods   143
    6.1   Acidic and Neutral Solutions   144
    6.2   Alkaline Solutions   150
    References   152

About the Authors   155

About the Series Editor   157

Index   159

Chapter 1
Introduction General Information on Technetium

In March 1869, the Russian scientist Dmitri Mendeleev presented to the Russian Society of Physics and Chemistry, the idea of the so-called periodic table. He proposed that chemical elements can be arranged in a table according to their periodically varying chemical properties. The significance of this discovery lies in the fact that the periodic table not only delivers information on chemistry of already known elements but also allows predicting the existence of not yet discovered elements with respective chemical properties. Mendeleev published his early version of the periodic table in a textbook for his students, entitled *The Principles of Chemistry* (Mendeleev 1869a, b), and in scientific journals: *Zhurnal Russkago Fiziko-Khimicheskago Obshchestva* and *Zeitschrift für Chemie* (Mendelejeff 1869). At that time, only around 60 elements were known and the announcement of Mendeleev's method of ordering of the chemical elements has stimulated researchers to search for new, undiscovered elements, the existence and properties of which were predicted by the periodic table.

One of the vacant places in the original periodic table of Mendeleev was located below manganese, indicating the existence of an unknown element with chemical properties similar to Mn and called eka-manganese. Kern (1877) was the first who announced the discovery of a new element whose chemical properties resembled those of manganese and which was supposed to be its analog. He named it *davyum*, in honor of the great British chemist Davy. Eventually, the material separated by Kern turned out to be an alloy of iridium, rhodium and iron. The *lucium* reported by Barriére (1896) also proved to be an already known element, in this case yttrium. Another attempt to discover the element 43 was made by Ogawa (see: Yoshihara 2008), who isolated from thorianite and molybdenum a material named *nipponium*, after Japan. This material also turned out to be not technetium, but most likely rhenium. The discovery of rhenium, however, is today attributed to Noddack et al. (1925). In 1925, they reported the discovery of two new elements present in the minerals of columbite, one of them was named rhenium while the other one, the "eka-manganese," was named masurium (after Masuria in the East Prussia, now Poland). These elements with the atomic numbers 75 and 43, respectively, filled

© Springer Nature Switzerland AG 2021
M. Chotkowski and A. Czerwiński, *Electrochemistry of Technetium*,
Monographs in Electrochemistry, https://doi.org/10.1007/978-3-030-62863-5_1

vacant places in the periodic table. A chemical analysis revealed that the columbite sample contained about 0.5 wt% of the newly discovered rhenium. The presence of the sought-after eka-manganese was deduced from the recorded X-ray lines: K1: 0.672 Å, K2: 0.675 Å and K1: 0.601 Å. Using a relation between X-ray frequencies of atoms and their atomic number presented by Moseley (1914), they identified lines of elements 43 and 75. However, other authors were unable to obtain the same results when repeating the measurements reported by Noddack et al. and the reported discovery of the technetium by the latter authors could not be confirmed. Therefore, the discussion about history of the discovery of the technetium could not be settled even until today (see: Armstrong 2003; Zingales 2006).

Today, Carlo Perrier and Emilio Segrè are credited with the discovery of the element with atomic number of 43. In 1937, they announced the discovery of radioactive isotopes of the new element in a sample prepared by the E. O. Lawrence (Perrier and Segrè 1937). The sample was a molybdenum plate subjected to several months of bombardment with deuterium nuclei (and secondary neutrons) in the University of California cyclotron in 1936. The cyclotron experiments were completed by the end of December 1936, and the irradiated plate was sent to Perrier at the University of Palermo in Italy, where in 1937 Perrier and Segrè detected strong ionizing radiation identified as slow electrons. In the first published study, Cacciapuoti and Segrè (1937) reported three radioactive species present in the analyzed plate, with only one of them being active enough to determine its half-life precisely. Today we know that this was ^{97m}Tc. In the next publication, Cacciapuoti (1939) reported ^{95m}Tc as the second discovered radioisotope of the Element 43.

Shortly after the discovery of the new element, the name *panorama* was proposed, after Panormus, a Latin name for Palermo, but this was not accepted. The new element is named *technetium* after the Greek word technètos, which means "artificial." The name was officially accepted by the 15th Conference of Pure and Applied Chemistry in 1949.

Measurable amounts of technetium were obtained for the first time in 1946 in the Oak Ridge reactor (USA) after neutron irradiation of a plate containing 4 kg of molybdenum. After completing the irradiation, the plate was subjected to a chemical treatment and, subsequently, 0.1 mg of metallic technetium were galvanically deposited on a copper foil (see: Enghag 2004).

Technetium is generated in stellar nucleosynthesis and its presence in the atmosphere of stars has been confirmed by means of analysis of stellar spectra (Merrill 1952). The TcI spectral lines are particularly intense for s-type stars. The technetium nucleosynthesis takes place in the star interior and the so formed Tc nuclei/atoms are transported toward the star surface. When the rate of the transport process is faster than or comparable to the decay rate of the longest living technetium isotope, i.e., ^{99}Tc, the latter nuclei may reach the star surface before they completely decay (Burbidge et al. 1957). This explains why Tc lines are present in the stellar spectra. The ^{99}Tc of stellar origin was most likely present in primordial material from which the Earth was formed. The half-life of this isotope ($2.11 \cdot 10^5$ y) is relatively short as compared with the age of the Earth ($4.5 \cdot 10^9$ y (Zhang 2002)) and all primordial ^{99}Tc incorporated into the Earth crust at the time of its formation is expected to have completely decayed since then. However, reliable measurements indicate the presence of nonman-made ^{99}Tc in Earth materials today. Traces of natural technetium were found in African pitchblende (Kenna 1964). The concentration of its long-lived isotope, ^{99}Tc, was determined at the level of $(2.5-3.1) \cdot 10^{-10}$ g per kg of the uranium ore. Clearly, naturally occurring ^{99}Tc present today in the Earth's crust must be generated in other than stellar nucleosynthesis processes, e.g., by spontaneous fission (SF) of naturally occurring uranium. The half-life of this process is equal to $1.5 \cdot 10^{16}$ y, $1.0 \cdot 10^{19}$ y and $8.2 \cdot 10^{15}$ y for [234]U, [235]U and [238]U, respectively (Holden and Hoffman 2000). The SF taking place in 1 kg of natural metallic uranium leads to a low but measurable emission of 13.5 neutrons per second (Reilly 1991). It is reasonable to assume that the naturally occurring terrestrial ^{99}Tc is produced mainly by spontaneous fission of [238]U. Knowing the SF cross-section of uranium and the probability of Tc generation in this process, one may calculate the amount of ^{99}Tc nuclei produced as a result of spontaneous fission of a given amount of uranium, e.g., present in a uranium-rich ore (Kuroda and Menos 1961; Kenna 1962; Parker and Kuroda 1958). Thus, one may use Eq. (1.1). to calculate the amount of ^{99}Tc present in the sample. N_{U-238}, N_{Tc-99} are the numbers of [238]U and ^{99}Tc nuclei, respectively. $\lambda_{U-238(fission)}$ is the spontaneous fission decay constant of [238]U (equal to $2.8 \cdot 10^{-24}$ s^{-1}) and λ_{Tc-99} is the decay constant of [99]Tc ($1.04 \cdot 10^{-13}$ s^{-1}). y_{Tc-99} denotes a fission yield of ^{99}Tc and is equal to 6.12% (Katakura 2012, Katakura et al. 2016 Nuclear Data Center).

$$y_{Tc-99}N_{U-238}\lambda_{U-238(fission)} = N_{Tc-99}\lambda_{Tc-99} \tag{1.1}$$

Using Eq. (1.1) and the estimated uranium content in the oceanic waters of the entire globe (Seko et al. 2003), which is equal to approximately 4.5 billion tons, one may calculate that the Earth oceans contain a few kg of ^{99}Tc of natural origin.

More than 30 isotopes of Tc are known today (see e.g., Baum 2010; Nystrom and Thoennessen 2012). All of them are radioactive with half-lives ranging from milliseconds (e.g., [112]Tc) to millions of years ([98]Tc). It must be stressed that the half-life of some of Tc isotopes has not yet been determined with acceptable reliability. The history of the discovery of known technetium isotopes was reviewed by Nystrom and Thoennessen (2012). The instability of the technetium nucleus was predicted by isobar rule formulated in 1934 by Josef Mattauch. It states that at least one of isobars of adjacent elements, e.g., [98]Mo and [98]Tc, must be unstable. The stable isotopes of the Tc neighbor with atomic number 44, i.e., ruthenium, are those with mass numbers of 96, 98–102 and 104 (Baum et al. 2010), while for another Tc neighbor, molybdenum (element 42), the stability is exhibited by the isotopes 92, 94–98 and 100. Therefore, the technetium isotopes with mass numbers from 94 to 102 are not stable (Johnstone 2017). The instability of the technetium nuclei is related to their nuclear shell structures (Mayer et al. 1951). The decay data for selected Tc radionuclides are summarized in Table 1.1.

Table 1.1 Properties of selected technetium radioisotopes (Browne 1986; Chu et al. 1999; Kobayashi 1998; Maiti 2010; Nystrom and Thoennessen 2012)

Isotope	Decay mode	Half-life	E_{max}^{β}/keV (intensity in %)	Principal X or gamma rays energy/keV (efficiency in %)	Generation
^{95}Tc	EC + β$^+$	20.0 h	925.2 (93.8)	X: 17.37 (19.4); 17.48 (37.1) γ: 765.8 (93.8)	nat.Zr(^{7}Li,xn) 94,94 m,95,96Tc nat.Mo(p,xn) 94/98Tc
^{95m}Tc	EC (95.7) IT (3.9) β$^+$ (0.4)	61 d	690.6 (30.1) 943.7 (38.0) 1729.9 (13.3)	X: 17.37 (19.1); 17.48 (36.4) γ: 204.1 (63.2); 511$_{ah}$ (0.8); 582.1 (29.9); 835.1 (26.6);	
^{96}Tc	EC + β$^+$	4.28 d	217.8 (19.5) 532.2 (79)	X: 17.37 (19.2) 17.48 (36.6) γ: 778.2 (100); 812.6 (82), 849.9 (98) 1126.9 (15.2)	
^{96m}Tc	IT (98.0) EC + β$^+$ (2.0)	51.5 min	1028.8 (1.5)	X: 18.25 (9.5); 18.37 (18.1); 20.599 (1.5); 20.62 (2.8)	nat.Zr(^{7}Li,xn) 94,94 m,95,96Tc
^{97}Tc	EC	2.6 · 10^6 y		X: 17.4; 18.3; 19.6	^{97}Mo(d,2n) ^{97}Tc
^{97m}Tc	IT (100) EC (<0.34)	91 d		X: 18.25 (14.2); 18.37 (27.0); 20.60 (2.2); 20.62 (4.2) γ: 96.5 (0.31)	^{96}Mo(d,n)^{97m}Tc
^{98}Tc	β$^-$	4.2 · 10^6 y	398.2 (100)	γ: 652.4 (100); 745.4 (102.7);	^{98}Mo(p,n)^{98}Tc
^{99}Tc	β$^-$	2.11 · 10^5 y	293.7 (100)	X: 19.15 (0.0002) γ: 89.65 (0.0006)	Fission, ^{99}Mo
^{99m}Tc	IT (99.9963) β$^-$(0.0037)	6.01 h	346.7 (0.0026) 436.4 (0.0010)	X: 18.25 (2.15); 18.37 (4.10) γ: 140.5 (89); 142.6 (0.019)	Fission, ^{99}Mo
^{100}Tc	β$^-$	15.8 s	3202.4 (100)	γ: 539.5 (7); 590.8 (5.7);	Fission,
^{101}Tc	β$^-$	14.22 min	1067.9 (6.4); 1306.1 (90.3)	γ: 306.9 (89); 545 (5.96);	Fission,
^{102}Tc	β$^-$	5.28 s	4054.9(3.5) 4530 (92.9)	γ: 475.1 (7)	Fission,

Table 1.2 Production yield of technetium isotopes as fission products, σ-cross section in barns / b (Katakura 2012; Katakura et al. 2016 Nuclear Data Center)

Isotope	$E_{neutron}$ ($\sigma_{(n,fission)}$ / b)	Production yield of fission products/%					
		Tc-97	Tc-98	Tc-99	Tc-100	Tc-101	Tc-102
U-235	25 meV (585.1)	$1.8 \cdot 10^{-11}$	$8.9 \cdot 10^{-7}$	6.14	$5.6 \cdot 10^{-6}$	5.19	4.18
	14 MeV (2.053)	$2.26 \cdot 10^{-7}$	$5.57 \cdot 10^{-6}$	5.16	$8.62 \cdot 10^{-4}$	3.46	3.23
U-238	14 MeV (1.136)	$2.47 \cdot 10^{-10}$	$1.40 \cdot 10^{-8}$	5.75	$1.13 \cdot 10^{-5}$	5.71	4.60
Pu-239	25 meV (747)	$8.16 \cdot 10^{-9}$	$4.17 \cdot 10^{-7}$	6.22	$3.07 \cdot 10^{-4}$	6.01	6.07
	14 MeV (2.334)	$3.32 \cdot 10^{-6}$	$6.08 \cdot 10^{-5}$	4.78	$6.71 \cdot 10^{-3}$	5.02	5.33

Nowadays, nuclear reactors are the main source of technetium isotopes. The uranium and plutonium nuclei fission leads to formation of the Tc isotopes with yields depending on the energy of incident neutrons (Table 1.2). ^{99}Tc is the only long-lived technetium isotope, which is generated in kilogram amounts. Despite large production yields of ^{101}Tc and ^{102}Tc, these isotopes have no practical importance due to their instability.

Technetium has physical and chemical properties typical for metals (Table 1.3). As compared with other d-block elements, Tc exhibits a relatively high melting point and an average density. The crystal lattice parameters are similar to rhenium. The ^{99}Tc nucleus has a nonzero spin ($I = 9/2$) and can be detected using NMR (see: Franklin et al. 1982; Tarasov et al. 2001; Poineau et al. 2010).

Interestingly, metallic technetium exhibits one of the highest superconductivity transition temperatures among those known for elements (Compton et al. 1961). The structure of the technetium compounds can be resolved using synchrotron radiation. It is worth noting that analysis of mixtures containing Tc and other metals, by means of, e.g., EXAFS, may be complicated by overlapping of some of X-ray absorption lines. For example, the K_{edge} of Tc (21.044 keV) is relatively close to L_{Iedges} characteristic of U (21.766 keV) or Np (22.438 keV) (Deslattes et al. 2003).

The electrochemical properties of metallic technetium are discussed in detail in Chap. 5. At this point, it is worth to mention that this element and its alloys have been intensively studied in the context of their conductivity. According to Koch and Love (1967) the resistivity of the metallic technetium varies nonlinearly with temperature. For temperatures higher than 77 K, this parameter (given in $\mu\Omega \cdot cm$) can be calculated with an accuracy greater than the nominal data error (±3–4%) using Eq. (1.2):

$$\rho = -3.191 + 7.844 \cdot 10^{-2}T - 2.816 \cdot 10^{-5}T^2 + 4.038 \cdot 10^{-9}T^3 \qquad (1.2)$$

At 25 °C ρ is equal to 185.0 n$\Omega \cdot$m.

Inorganic salts and organic compounds of technetium are typical raw materials used in electrochemical studies. Their solubilities were discussed by Rard et al. (1999). Their values are presented in Table 1.4.

Table 1.3 Selected physical and chemical properties of technetium (Meggers and Scribner 1950; Murin et al. 1961; Rard et al. 1999; Thomson et al. 2001; Stone 2005; Shibata et al. 2011)

Property	Value	
Atomic number	43	
Atomic weight	98.8	
Electronic structure and of the shell (ground state)	$[Kr] 5s^1 4d^6 (^6D_{9/2})$	
Density / g cm^{-3}	11.487	
Melting point / °C	2 157	
Boiling point / °C	4 265	
Atomic radius / Å	1.358	
Parameters of the crystal lattice / Å	a = 2.741	
	c = 4.399	
NMR of Tc-99		
Magnetic momentum / μ_Z	5.6847	
Electric quadrupole moment / $10^{-28} m^2$	−0.129	
Magnetogiric ratio, γ / 10^7 rad s^{-1} T^{-1}	−6.0503	
Lines of the optical spectrum / Å (rel. intensity)	4297.06 (500)	
	4262.26 (400)	
	4238.19 (300)	
	4031.63 (300)	
	3636.10 (400)	
	3466.29 (250)	
Energies of X-ray emission lines / eV (rel. intensity)		
$K_{\alpha1}$	18 367.1 (100)	
$K_{\alpha2}$	18 250.8 (53)	
$K_{\beta1}$	20 619 (16)	
$L_{\alpha1}$	2 424 (100)	
X-ray absorption lines / eV		
Edge K	21 044	
Edge L_I	3 043	
Edge L_{II}	2 793	
Edge L_{III}	2 677	
Cross section of ^{99}Tc	For 25.3 meV	For 14 MeV
Total	28.26 b	4.165 b
Elastic	4.657 b	2.178 b
Capture	23.60 b	9.255 mb
Activity	100 MBq ~ 0.158 g ^{99}Tc or 0.51 ng ^{99m}Tc	

Among the compounds listed in Table 1.4, the highest solubility is exhibited by the sodium salts. Ph_4AsTcO_4, on the other hand, is considered as almost insoluble and was frequently used in gravimetric analysis of the technetium. Solubility of tetralkylammonium pertechnetates is practically independent on the ionic strength of the solution, as reported by Konstantin German and coworkers (German et al. 1988). They also referred to an earlier work of Spitsyn who showed that the solubility

Table 1.4 Solubilities of pertechnetate salts in water (Keller and Kanellakopulos 1963; Busey and Bevan 1960; Boyd 1978; German et al. 1988, 1993; Könnecke et al. 1997; Peretrukhin et al. 2008)

Salt	Solubility	Temperature/°C
NH_4TcO_4	0.594 mol kg^{-1}	25
$NaTcO_4 \cdot 4H_2O$	11.299 mol kg^{-1}	
$KTcO_4$	0.1057 mol kg^{-1}	
$CsTcO_4$	0.0184 mol kg^{-1}	
$AgTcO_4$	0.0270 mol kg^{-1}	
$TlTcO_4$	0.00248 mol kg^{-1}	
$RbTcO_4$	0.04699 mol dm^{-3}	20
$Ba(TcO_4)_2$	0.165 mol dm^{-3}	
$(CH_3)_4NTcO_4$	0.135 mol dm^{-3}	25
$(C_2H_5)_4NTcO_4$	0.025 mol dm^{-3}	
$(n-C_4H_9)_4NTcO_4$	0.0043 mol dm^{-3}	
Ph_4AsTcO_4	$4.0 \cdot 10^{-4}$ mol dm^{-3}	

of $Sr(TcO_4)_2$ is lower as compared with $Ca(TcO_4)_2$. Solubility of the latter can be estimated at a level of 1 mol dm^{-3}.

References

Armstrong JT (2003) Technetium. https://pubsapp.acs.org/cen/80th/technetium.html? Accessed 13 Apr 2018

Barriére (1896) Luminous material for incandescent gas-lighting. US Patent 567571A, 8 Sept 1896

Baum E et al (2010) Nuclides and isotopes chart of the nuclides, 17th edn. Bechtel

Boyd GE (1978) Osmotic and activity coefficients of aqueous $HTcO_4$ and $HReO_4$ solutions at 25 °C. J Solution Chem 7:229–238

Browne E, Firestone RB (1986) Table of radioactive isotopes. Wiley, USA

Burbidge KM, Burbidge GR, Fowler GA et al (1957) Synthesis of the elements in stars. Rev Mod Phys 29(4):547–654

Busey RH, Bevan Jr RB (1960) Chemistry of technetium, solubility and heat of solution of potassium pertechnetate and potassium hexachlororhenate(IV). Tech Report ORNL-2983, Oak Ridge National Laboratory, Tennesse

Cacciapuoti BN, Segré E (1937) Radioactive isotopes of element 43. Phys Rev 52:1252–1253

Cacciapuoti BN (1939) Radioactive isotopes of element 43. Phys Rev 55:110

Chu SYF, Ekström LP, Firestone RB (1999) WWW table of radioactive isotopes, database version 1999-02-28 from URL nucleardata.nuclear.lu.se/toi/index.asp

Compton VB, Corenzwit E, Maita JP (1961) Superconductivity of technetium alloys and compounds. Phys Rev 123:1567–1568

Deslattes RD, Kessler EG Jr, Indelicato P (2003) X-ray transition energies database. Rev Mod Phys 75:35–99

Enghag P (2004) Encyclopedia of the elements: technical data—history—processing—applications, 1st edn. Wiley-Vchy, p 652

Franklin KJ, Lock CJL, Sayer BG (1982) Chemical applications of ^{99}Tc NMR spectroscopy: preparation of novel Tc(VII) species and their characterization by multinuclear NMR spectroscopy. J Am Chem Soc 104:5303–5306

German KE, Krjuchkov SV, Belyaeva LI et al (1988) Ion association in tetraalkylammonium pertechnetate solutions. J Radioanal Nucl Chem 121(2):515–521

German KE, Grushevschkaya LN, Kryutchkov SV et al (1993) Investigation of phase transitions and other physico-chemical properties of pertechnetates and perrhenates of alkali and organic cations. Radiochim Acta 63:221–224

Holden NE, Hoffman DC (2000) Spontaneous fission half-lives for ground-state nuclides (Technical Report). Pure Appl Chem 72(8):1525–1562

Johnstone EV et al (2017) Technetium: the first radioelement on the periodic table. J Chem Edu. 94(3):320–326

Katakura J (2012) JENDL FP decay data file 2011 and fission yields data file 2011. JAEA-Data/Code 2011-025, Nuclear Data Center. https://wwwndc.jaea.go.jp/cgi-bin/FPYfig?lib=11&mode=s&iso=sU238&typ=g1&ynuc=Tc-99&ysub=Show+Yield&yld=c&zlog=set&ylow=12&xpar=a&eng=e1. Accessed 13 Apr 2018

Katakura J, Minato F, Ohgama K (2016) Revision of the JENDL FP fission yield data. In: EPJ web of conferences, vol 111, p 08004

Keller C, Kanellakopulos B (1963) Darstellung und Untersuchung einiger Pertechnetate des Typs Me^{I}TcO$_4$. Radiochim Acta 1:107–108

Kenna BT, Kuroda PK (1961) Isolation of natural occurring technetium. J Inorg Nucl Chem 23:142–144

Kenna BT (1962) The search for technetium in nature. J Chem Educ 39(9):436–442

Kenna BT, Kuroda PK (1964) Technetium in nature. J Inorg Nucl Chem 26(4):493–499

Kern S (1877) On a new metal, Davyum. The London, Edinburgh, and Dublin Philos Mag J Sci Series 5. 4(23):158–159

Kobayashi T et al (1998) Decay properties of 97m'gTc. Nucl Phys A 636:367–378

Koch CC, Love GR (1967) The electrical resistivity of technetium from 8.0° to 1700° K. J Less Comm Met 12(1):29–35

Könnecke Th, Neck V, Fanghänel Th, Kim JI (1997) Activity coefficients and pitzer parameters in the systems Na$^+$/Cs$^+$/Cl$^-$/TcO$_4^-$ or ClO$_4^-$/H$_2$O at 25^{0}C. J Solution Chem 26(6):561–577

Kuroda PK, Menos M (1961) Determination of trace quantities of fission products in nonirradiated natural and depleted uranium salts. Nucl Sci Eng 10:70

Maiti M et al (2010) Separation of no-carrier-added 93,94,94m,95,96Tc from 7Li induced natural Zr target by liquid–liquid extraction. Appl Radiat Isot 68:42–46

Mauttach J (1934) Zur Systematik der Isotopen. J Z Physik 91:361–371

Mayer MG, Moszkowski SA, Nordhem LW (1951) Nuclear shell structure and beta-decay I. Odd a Nuclei. Rev Modern Phys 23(4):315–321

Meggers WF, Scribner BK (1950) Arc and spark spectra of technetium. J Res Natl Bur Stand 45(6):476–489

Mendeleev DI (1869) Osnovy Khimii [The Principles of Chemistry]. St. Petersburg

Mendeleev DI (1869) Sootnoshenie svoistv s atomnym vesom elementov (The correlation of properties with the atomic weights of the elements). Zh Russ Fiz-Khim O-va 1:60–77

Mendelejeff D (1869) Über die Beziehungen der Eigeschaften zu den Atomgewichten der Elemente (On the relationship of the properties of the elements to their atomic weights). J Z Chem 12:405–406

Merrill PW (1952) Technetium in the stars. Science 115(2992):479–489

Moseley HGJ (1913) The high-frequency spectra of the elements. Phil Mag 26(13):1024–1034

Moseley HGJ (1914) The high-frequency spectra of the elements. Part II Phil Mag 27:703–713

Murin AN, Nefedov VD, Ryukhin YA (1961) Technetium—element 43. Russ Chem Rev 30:107–115

Noddack W, Tacke I, Berg O (1925) Die Ekamangane. Naturwissenschaften. 13:567–574

Nystrom A, Thoennessen M (2012) Discovery of yttrium, zirconium, niobium, technetium and ruthenium isotopes. At Data Nucl Data Tables 98:95–119

Parker P, Kuroda PK (1958) The occurrence of molybdenum-99 in natural and in depleted uranium salts and the spontaneous fission half-life of uranium-238. J Inorg Nucl Chem 5:153–158

Peretrukhin VF, Moisy F, Maslennikova AG et al (2008) Physicochemical behavior of uranium and technetium in some new stages of the nuclear fuel cycle. Rus J General Chem 78(5):1031–1046

Perrier C, Segrè E (1937) Some chemical properties of element 43. J Chem Phys 5:712–716

Poineau F, Weck PF, German K (2010) Speciation of heptavalent technetium in sulfuric acid: structural and spectroscopic studies. Dalton Trans 39:8616–8619

Rard JA, Rand MH, Anderegg G et al (1999) Chemical thermodynamics of technetium, vol 3. Elsevier

Reilly D, Ensslin N, Hastings S Jr, Kreiner S (eds) (1991) Passive nondestructive assay of nuclear materials. NUREG/CR 5550, LA-UR-90–732, Office of Nuclear Regulatory Research, US Nuclear Regulatory Commission, Washington, DC 20555, p 413. ISBN 0–16–032724–5

Seko N, Katakai A, Hasegawa S et al (2003) Aquaculture of uranium in seawater by a fabric-adsorbent submerged system. J Nucl Technol 144:274–278

Shibata K, Iwamoto O, Nakagawa T et al (2011) JENDL-4.0: A New Library for Nuclear Science and Engineering. J Nucl Sci Technol 48(1):1–30. JENDL-4.0. https://wwwndc.jaea.go.jp/jendl/j40/j40.html. Accessed 20 Apr 2019

Stone NJ (2005) Table of nuclear magnetic dipole and electric quadrupole moments. Atomic Data Nucl Data Tables 90:75–176

Tarasov VP, Muravlev YuB, German KE et al (2001) ^{99}Tc NMR of supported technetium nanoparticles. Dok Phys Chem 377(1–3):71–76

Thomson AC, Attwood DT, Gullikson Eric M et al (2001) X-ray data booklet center for X-ray optics and advanced light source, Lawrence Berkeley National Laboratory. https://www.psi.ch/sites/default/files/import/sls/ms/Station1_DescriptionEN/X-ray_data_booklet.pdf. Accessed 12 Nov 2019

Yoshihara HK (2008) Nipponium as a new element (Z = 75) separated by the Japanese chemist, Masataka Ogawa: a scientific and science historical re-evaluation. Proc Jpn Acad Ser B Phys Biol Sci 84(7):232–245

Zhang Y (2002) The age and accretion of the earth. Earth Sci Rev 59:235–263

Zingales R (2006) The history of element 43—technetium (the author replies). J Chem Edu 83(2):213

Chapter 2
Comprehensive Electrochemistry of Tc and Its Periodic Table Neighbors

Technetium is located in group 7 of the periodic table together with manganese and rhenium. When elements of the periods of the periodic table are analyzed, one finds technetium in the period 5 surrounded with molybdenum and ruthenium as its neighbors (Fig. 2.1). Being transition metal technetium has seven valence electrons and these electrons reside in the 5s and 4d subshells. These seven electrons located outside of the krypton's noble gas core can participate in the formation of chemical bonds. As a result, the technetium forms compounds with oxidation states varying from -1 to $+7$.

Although many aspects of technetium chemistry are similar to what is observed for all its neighbors from the periodic table, its chemical properties most resemble those exhibited by rhenium. This phenomenon is attributed to the equal values of atomic radii of technetium and rhenium, which is a result of lanthanides contraction effects (see Table 2.1). Further on, the electronegativity of both elements is also very similar. Both elements have extensive and often similar oxohalides chemistry (Rard 1985; Rouschias 1974). Therefore, rhenium is most common considered as a technetium analog that can be used for deduction of properties of the latter.

Among chemical properties of Tc that are similar to those exhibited by Mo and Ru, the most notable is tendency to form polymeric species containing several Tc atoms. The polymeric species of ruthenium and molybdenium with $+3$ or $+4$ oxidation states (e.g., $[Ru_4(OH)_{12}]^{4+}$ or $[Mo_3O_4]^{3+}$) are well known (Ardon and Pernick 1974; Paffett and Anson 1983; Isherwood 1985) and their examples are listed in Table 2.1. The tendency to polymerization (or dimerization) is also characteristic for reduced technetium(III/IV). Well known are -Tc-O-Tc- species, e.g., $[Tc_2O_2]^{3/4+}$ or $[Tc_3O_5]^{2+}$, whose structures are similar to the respective Mo or Ru polymers (see e.g. Mausolf et al. 2011; Baumann et al. 2018).

According to Encyclopedia Britannica, the metals that are located in the VIIB, VIII, and IB groups (old CAS notation) of the second and third transition series (Tikkanen 2008) are considered as the noble ones. Although this classification is not recommended by IUPAC, but it is still common. Consequently, technetium and rhenium, which are located in group VIIB, are sometimes considered as noble metals

© Springer Nature Switzerland AG 2021
M. Chotkowski and A. Czerwiński, *Electrochemistry of Technetium*,
Monographs in Electrochemistry, https://doi.org/10.1007/978-3-030-62863-5_2

Fig. 2.1 Technetium is
surrounded by its periodic
table neighbors

	$_{25}$Mn	
$_{42}$Mo	$_{43}$Tc	$_{44}$Ru
	$_{75}$Re	

Table 2.1 Technetium and its neighbors' species in aqueous solutions and the standard redox potentials between their selected couples

		Mn	Tc	Re	Mo	Ru
Atomic radius of M/pm		140	135	135	145	130
Pauling electronegativity		1.55	1.9	1.9	2.16	2.22
$E^{\ominus}_{MO_4^- + 4H^+ + 3e^- \leftrightarrows MO_2 + 2H_2O}$		1.70	0.746	0.51		1.533
$E^{\ominus}_{MO_4^- + e^- \leftrightarrows MO_4^{2-}}$		0.56	−0.64	−0.7		0.593
$E^{\ominus}_{MO_2 + 4H^+ + 4e^- \leftrightarrows M + 2H_2O}$		0.025	0.272	0.276	−0.152	0.68
Selected forms of the elements in aqueous solutions	VII	MnO_4^-	TcO_4^-	ReO_4^-		RuO_4^-
	VI	MnO_4^{2-}	TcO_4^{2-}	ReO_3	MoO_4^{2-}	RuO_4^{2-}
			$TcO(OH)_3^-$			RuO_2^{2+}
	V	MnO_4^{3-}	TcO^{3+}	Re_2O_5		Ru_2O_5
	IV	MnO_2	TcO_2	ReO_2	MoO_2	RuO_2
			$[Tc_2O_2]^{4+}$		$Mo_3O_4^{4+}$	$[Ru_4(OH)_{12}]^{4+}$
	Mixed III/IV		$[Tc_2O_2]^{3+}$		$Mo_3O_4^{3+}$	
	III	Mn^{3+}	Tc^{3+}	Re^{3+}(?)	Mo^{3+}	Ru^{3+}
			TcO^+		possible	possible
			possible		polymer.	polymer. forms
			polymer.		forms	
			forms			

along with ruthenium. The standard redox potentials of MO_2/M(metal) redox couples of Re and Tc in acidic solutions are almost the same (Bard et al. 1985) (Table 2.1).

Cyclic voltammetry curves recorded on metallic Tc and Re electrodes in pure aqueous acidic or alkaline electrolytes do not reveal well developed the so called electric double layer charging region, i.e., the potential range free from faradaic reactions. The latter feature is characteristic of noble metals: Pt or Au.

Our knowledge on technetium chemistry in aqueous solutions has been significantly expanded since the first publication of potential pH diagram for selected technetium species by Pourbaix (e.g. de Zoubov and Pourbaix 1966). The first published such type of charts assumed Tc^{2+} ions as a form of Tc that is stable in acidic solutions at potentials near the hydrogen evolution region (most probably as an analogy to Mn^{2+}). It is well known that this form of Tc is extremely unstable and cannot

Fig. 2.2 E_h-pH diagram for technetium constructed using *Geochemist's Workbench* program. The shaded area represents the region in which the amorphous solid, $TcO_2 \cdot 2H_2O_{(am)}$ is stable. Concentration of Tc 10^{-8} mol dm^{-3} (reprinted with permission from Icenhower et al. (2010). Copyright 2010 American Journal of Science)

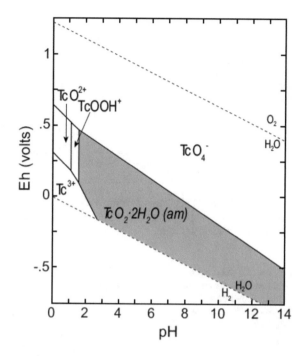

be observed in noncomplexing aqueous solution. Even nowadays an analysis of the stability regions of selected technetium species is difficult due to incomplete physicochemical data. Figure 2.2, for example, shows the potential pH diagram for Tc-O-H system for total Tc concentration of 10^{-8} mol dm^{-3} which was constructed using *Geochemist's Workbench* program.

Pertechnetates as well as perrhenates reveal thermodynamic stability over broad ranges of pH and potentials. Under reducing conditions, Tc(IV) is the preferred form of Tc. Its structure strongly depends on the solution acidity. In alkaline, neutral and weak acidic solutions, the Tc(IV) exists as TcO_2 (or $TcO(OH)_2$ in more diluted solutions) while $TcO(OH)^+$ or TcO^{2+} are characteristic of more concentrated acidic solutions.

Current knowledge on chemistry Tc(III) and Tc(IV) forms is also much broader than 50 years ago. Figure 2.2 reports Tc(III) as Tc^{3+} ions, which are observed at pH lower than 3. This behavior makes technetium similar to molybdenum and ruthenium because Mo^{3+} and Ru^{3+} ions are also stable at low pH. Figure 2.3 presents an example of the potential pH diagram for ruthenium. Unfortunately, even for technetium neighbors, the Pourbaix diagrams published by various authors are not consistent. Popov and Spinu (2016) published one of the most recent potential pH diagrams for Ru (Fig. 2.3). The reader can find an area characteristic of polymeric forms of Ru(IV) at pH near 3. Consequently, one may expect the existence of Tc(IV) polymers also for Tc at mmol dm^{-3} concentration. Vongsouthi (2009) concluded that these polymeric Tc forms should be observed at pH lower than about 1.7.

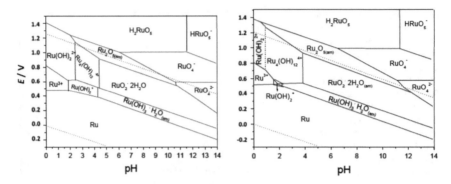

Fig. 2.3 The potential pH diagram for Ru-H$_2$O system, (left panel) $c_{Ru}^0 = 10^{-6}$ mol dm^{-3} (right panel) $c_{Ru}^0 = 10^{-4}$ mol dm^{-3} (reprinted with permission from Povar and Spinu (2016). Copyright 2016 Creative Common License)

Fig. 2.4 E_h-pH diagram obtained from the equations published by Pourbaix (reprinted with permission from Schrebler et al. (2001). Copyright 2001 Elsevier)

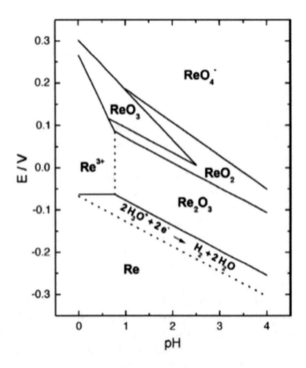

When the total concentration of ruthenium increases the area of stability of Ru(OH)$_2^{2+}$ ions becomes more narrow while for Ru polymeric species an opposite effect is observed. Other than ionic forms of Ru present in the diagrams include Ru$_2$O$_5$ and Ru(OH)$_3$. The existence of Ru(OH)$_3$ makes this element similar to rhenium for which Re$_2$O$_3$ is observed (Fig. 2.4). Another one aspect that is characteristic of Re is the presence of oxides with higher oxidation states, e.g., ReO$_3$, that is stable in acidic

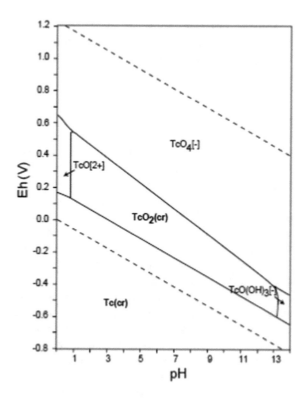

Fig. 2.5 E_h-pH diagram for technetium constructed using thermodynamic data published by OECD-NEA. Concentration of Tc 10^{-10} M. (reprinted with permission from Atlas of E_h-pH diagrams (2005). Copyright 2005 Creative Common License)

solutions. The existence and electrogeneration of analogous technetium species are discussed in detail in a later part of this chapter and in Chap. 3.

The technetium diagram shown in Fig. 2.2 does not include all Tc species observed in noncomplexing aqueous solutions. Warwick et al. (2007) discussed a slow dissolution of TcO_2 at pH higher than 13, which leads to the formation of $TcO(OH)_3^-$. This anionic form of Tc(IV) has been included in OECD-NEA E_h-pH diagram presented in Fig. 2.5. (See: Atlas of E_h-pH diagrams 2005.) In contrast to the diagram calculated using *Geochemist's Workbench* program, the OECD-NEA data report TcO^{2+} as the only ionic form of Tc(IV) present in acidic solutions. Noteworthy is the fact that the latter diagram includes also the stability region of metallic technetium. This region is separated from the TcO_2 field by a line parallel to the hydrogen evolution line and shifted by about 0.2 V in respect to the latter.

If the same type of the diagram is to be created for an extended range of hydrogen ions concentrations (from $-\log[H^+] < 0$ to $-\log[H^+] > 14$), one must consider the existence of additional Tc(VII) and Tc(V) (and Tc(VI)?) species. Poineau et al. (2018) reported a transformation of pertechnetates to TcO_3^+ ions in extremely concentrated acid solutions ($-\log[H^+] < 0$). The latter ions can undergo reduction to stable Tc(V) form (as TcO^{3+}). One may expect stabilization of technetium with higher oxidation states: Tc(V) (as $TcO(OH)_3^{2-}$?) or even Tc(VI) also in concentrated alkaline solutions ($-\log[H^+] > 14$).

Dioxides of technetium and its neighbors from the periodic table are well known and their chemistry is well recognized. Among them only manganese dioxide is a strongly oxidizing agent. The standard redox potential of MnO_2/Mn^{2+} couple in acidic solutions is high and equals 1.23 V. TcO_2 or ReO_2 is not considered as oxidants.

An analysis of the values of the standard redox potentials of $MO_4^-/M(IV)$ couples of the manganese group elements reveals that the values of these parameters decrease in the order from manganese through technetium to rhenium. In contrast to pertechnetates and perrhenates, the permanganates are strong oxidizing agents in acidified aqueous solutions. The values of the respective standard redox potentials for TcO_4^-/TcO_2 or ReO_4^-/ReO_2 systems are relatively low and equal to 0.746 V and 0.51 V, respectively. Similar periodic changes of the redox potentials are observed also for $MO_4^-/M(VI)$ couples. The electroreduction of MnO_4^- to stable in alkaline solution MnO_4^{2-} proceeds easily and quickly and its completion does not cause significant experimental problems (Freeman and Mamantov 1976; Norwell and Mamantov 1977). In the case of technetium, on the other hand, this process is more complicated due to instability of Tc(VI). The latter is generated in the first step of the pertechnetates electroreduction and rapidly disporoportionates to Tc(VII) and Tc(IV) via another one unstable intermediate, i.e. Tc(V) (Rard et al. 1999). It should be noted, however, that the latest report indicates possibility of extending the life time of Tc(VI) in alkaline environment with high ionic strength (Chatterjee et al. 2018).

What distinguishes technetium (and rhenium) from manganese is the lack of ionic form with +2 oxidation state that are stable in aqueous solutions. Mn^{2+} is especially stable in an aqueous environment and can be oxidized on solid electrodes to Mn^{3+} and further to MnO_2 (Chotkowski et al. 2011; Rogulski et al. 2006). Lee and coworkers performed electrochemical studies on properties of manganese dioxide (Lee et al. 1977, 1980). Both Mn^{3+} and Tc^{3+} ions are sensitive to the presence of oxygen, Tc(III) is quickly converted to Tc(IV) in a reaction with oxygen. On the basis of spectroelectrochemical results, Córdoba de Torresi and Gorenstein (1992) concluded that the first step of the MnO_2 reduction leads to the generation of MnO(OH). This reversible stage involves a one electron reduction of Mn(IV) accompanied by addition of a proton to MnO_2. The proton originates from decomposition of a water molecule, Eq. (2.1).

$$MnO_2 + H_2O + e^- \leftrightarrows MnO(OH) + OH^- \qquad (2.1)$$

MnO(OH) can be further reduced to a mixture of hydrated manganese oxides (II and III), $MnO_2\ Mn_3O_4$ that can be further reduced to $Mn(OH)_2$. Early papers devoted to electrochemical properties of Tc assume the existence of its mixed oxides, e.g., Tc_3O_4, and hydroxides, e.g., $Tc(OH)_2$ (Cartledge 1971; Mazzocchin et al. 1974). Contemporary reports, however, are not conclusive as to their existence and structure (Rard et al. 1999). Moreover, in contrast to technetium and rhenium, manganese is the only manganese group member have found broad applications in galvanic cells, including classical Leclanché batteries, their modifications and other power sources

Table 2.2 The half-wave potentials, reduction efficiency of the depolarizer, limiting ionic conductivity at 25 °C and diffusion coefficient of Mn, Tc and Re species refer to the first reduction step presented in Fig. 1.2 (Astheimer and Schwochau 1976)

Depolarizer	Solvent	Parameter				Valence state
		$\eta/\%$	$E_{1/2}/V$ versus SCE	Λ_0 $\Omega^{-1}\ cm^2\ val^{-1}$	$D_0\ /$ $10^{-5}\ cm^2\ s^{-1}$	
MnO_4^-	Water	13	0.01	63.5	1.69	II
	Acetonitrile	(88)	−0.60	(117.4)	(3.13)	VI
TcO_4^-	Water	10	−0.84	55.5	1.48	V
	Acetonitrile	100	−1.74	127.5	3.40	VI
	Dimethyl-sulfoxide	100	−1.86	26.2	0.70	VI
	Dimethyl-acetamide	–	−1.78	40.2	1.07	VI?
ReO_4^-	Water	<5	−1.60	54.8	1.46	V
	Acetonitrile	100	−2.30	127.5	3.40	VI
	Dimethyl-sulfoxide	100	−2.44	26.1	0.70	VI
	Dimethyl-acetamide	–	−2.35	40.0	1.06	VI?

(e.g. Crompton and Crompton 2000; Czerwiński and Rogulski 2005; Hamankiewicz et al. 2014).

The differences in the electroreduction process of permanganates, pertechnetates and perrhenates are presented in a very clear manner by Astheimer and Schwochau (1976). Table 2.2 presents the selected transport and electrodic properties of reduced ionic forms of manganese, technetium and rhenium. Mn(VI), Tc(VI) and Re(VI) were obtained in acetonitrile solutions containing ~40 mM $(CH_4)_4NClO_4$ as a supporting electrolyte by means of controlled potential electrolytic reduction of a respective depolarizer at potentials of $E = E_{1/2} − 0.2$ V.

When organic solvents, such as acetonitrile, dimethylsulfoxide or dimethylacetamide, are considered, one may found that diffusion coefficients or limiting conductivity of Re(VI) in such liquids are the same as for Tc(VI). More pronounced differences in values of these parameters measured for technetates(VI) and rhenates(VI) are observed for aqueous solutions (Table 2.2).

An analysis of the table shows that the products of the electroreduction of TcO_4^- and ReO_4^- have the same oxidation state and exhibit almost the same values of the diffusion coefficient and limiting conductivity. These similarities indicate that the structures of products of the electrochemical reduction of pertechnetates and perrhenates should be the same. In general, it can be said that if a reduction process of rhenium species exists then a respective process most probably takes place also for the technetium. The corresponding redox potentials of Re couples will be lower than those determined for respective Tc analogs. If the electrode potential of selected redox Tc system is very negative (close to the electrolyte stability limit) it may be impossible to determine redox potentials of the respective rhenium compounds due to the electrochemical decomposition of the medium.

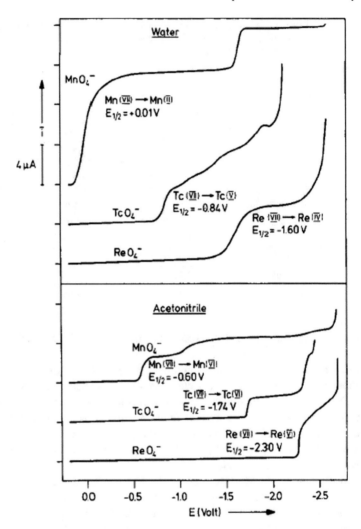

Fig. 2.6 D.C. polarograms of the tetramethylammonium salts of 10^{-3} M MnO_4^-, TcO_4^- and ReO_4^- in water and acetonitrile. Supporting electrolyte 0.05 M $(CH_3)_4NClO_4$, drop time 0.5 s, flow rate of mercury 0.923 mg s^{-1}, temperature 25 °C (reprinted with permission from Astheimer and Schwochau (1976). Copyright 1976 Elsevier)

Astheimer and Schwochau (1976) have shown that reduction of MO_4^- (M: Mn, Tc, Re) in various nonaqueous solutions may lead to the formation of ionic forms of the elements with the same oxidation state and almost the same, at least for Tc and Re, transport properties. In aqueous media, they obtained Tc(V), Re(IV) and Mn(II) species. Respective polarograms for discussed Mn, Tc and Re systems are presented in Fig. 2.6. The curves recorded for technetium in aqueous solutions reveal a poorly

Fig. 2.7 Cyclic voltammograms of 1.47 mM [Re(dmpe)$_2$Cl$_2$]$^+$ (—) and 1.62 mM [Tc(dmpe)$_2$Cl$_2$]$^+$ (- - -) in 0.1 M SDS/0.1 M TEAP/H$_2$O. The scan rate is 100 mV s^{-1} (reprinted with permission from Kirchhoff et al. (1988). Copyright 1988 American Chemical Society)

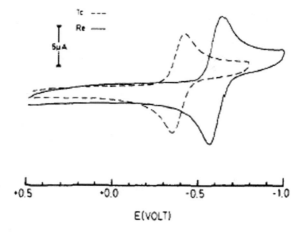

shaped signal at potentials lower than -1 V. This indicates that the reduction of Tc is completed when species with lower than $+$V oxidation states are generated.

A series of works by Edward Deutsch and William Heinemann with coworkers (e.g. Kirchhoff et al. 1987, 1988) made a significant contribution to comprehensive electrochemistry of technetium and rhenium compounds in aqueous and nonaqueous solutions. Figure 2.7 presents a cyclic voltammogram recorded for a solution containing *trans*–[Tc(or Re)D$_2$X$_2$]$^{+/0}$ complexes where X represents a halide ion while D states for an organic ligand.

A characteristic feature of the voltammetric curves recorded for *trans*–[Tc(or Re)D$_2$X$_2$]$^{+/0}$ redox couples is the shift of half waves toward more positive values when technetium replaces rhenium. This difference between $E_{1/2}$ values for Tc and Re is equal to approximately 0.2–0.22 V and, as follows from Table 2.3, only weakly depends on identity of organic and inorganic ligand. Kirchhoff et al. (1987) analyzed $E^{0\prime}$(Tc) as a function of $E^{0\prime}$(Re) for *trans*–[Tc(or Re)D$_2$X$_2$]$^{+/0}$ complexes containing Tc(or Re)(III/II) and Tc(or Re)(II/I) and a tertiary phosphine or arsine ligand indicated here by D. In DMF, this relationship turned out to be linear with a slope of 1.04 ± 0.01 and an intercept of 219 ± 15 mV. Results of additional spectroscopic measurements were found to be in line with the electrochemical results. Based on absorption maxima of Tc(III) or Re(III) complexes, these authors noted that for given rhenium complexes a charge-transfer process (HTMCT transition) occurs at an energy ca. 260 ± 30 mV (2140 ± 270 cm^{-1}) higher than for the respective technetium complexes. This confirms that the latter are easier to reduce than the respective rhenium species.

A comparison of $E_{1/2}$ values obtained for selected Tc and Re redox couples shows that they differ usually by 0.2–03 V although some significant deviations from this rule are noticeable. For example Tisato et al. (1990) examined Tc^V/Tc^{IV} and Re^V/Re^{IV} redox systems containing MO^{3+} cores and polidentate Schiff bases as the ligands. The electroreduction of Tc(V)-oxocomplexes occurred at a potential almost 0.5 V higher than for the respective rhenium analogs.

Table 2.3 Redox potentials for the *trans*–[Tc(or Re)D$_2$X$_2$]$^{+/0}$ couples (Ichimura et al. 1984; Kirchhoff et al. 1987, 1988)

Solution	Redox couple	$E_{1/2}$ versus Ag,AgCl (3 M KCl)
0.1 M SDS + 0.1 M TEAP in H$_2$O	[Tc(dmpe)$_2$Cl$_2$]$^{+/0}$	−0.396
	[Re(dmpe)$_2$Cl$_2$]$^{+/0}$	−0.600
	[Tc(dmpe)$_2$Br$_2$]$^{+/0}$	−0.300
	[Re(dmpe)$_2$Br$_2$]$^{+/0}$	−0.504
	[Tc(depe)$_2$Cl$_2$]$^{+/0}$	−0.404
	[Re(depe)$_2$Cl$_2$]$^{+/0}$	−0.592
0.5 M TEAP in DMF	[Tc(diars)$_2$Cl$_2$]$^{+/0}$	−0.061
	[Re(diars)$_2$Cl$_2$]$^{+/0}$	−0.319
	[Tc(dppe)$_2$Cl$_2$]$^{+/0}$	−0.010
	[Re(dppe)$_2$Cl$_2$]$^{+/0}$	−0.205
	[Tc(dmpe)$_2$Br$_2$]$^{+/0}$	−0.098
	[Re(dmpe)$_2$Br$_2$]$^{+/0}$	−0.297

dmpe: 1,2-bis(dimethylphosphino)ethane
depe: 1,2-bis(diethylphosphino)ethane
diars = 1,2-bis(dimethylarsino)benzene
dppe = 1,2-bis(diphenylphosphino)ethane

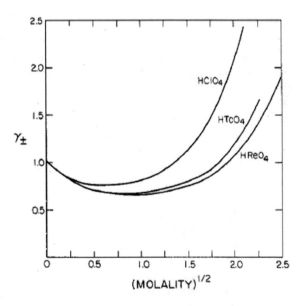

Fig. 2.8 Concentration dependence of the mean molal ionic activity coefficients for aqueous binary solutions of HClO$_4$, HTcO$_4$, and HReO$_4$ at 25 °C (Boyd 1978, reprinted with permission from Boyd (1978). Copyright 1978 American Chemical Society)

Boyd (1978) determined the activity coefficient, $\gamma_{\pm}$, of HTcO$_4$ and HReO$_4$ acids. Figure 2.8 shows that the transport properties of the technetium species are almost the same as for the respective rhenium compounds and this effect is observed for both aqueous and nonaqueous environment. This conclusion is in line with the results reported by Astheimer and Schwochau (1976). The relative difference in the respective values calculated for solutions of both acids does not exceed 3% for concentrations of up to 1.0 mol kg^{-1} (Boyd 1978). Thus, one may deduce transport parameters for Tc on the basis of respective data available for Re. Such an approach is justified and does not generate large errors as long as rhenium solutions with millimolar concentrations are considered.

Ironically, what makes the electrochemical properties of technetium and rhenium similar in aqueous solutions is their incomplete description and understanding.

The electrochemical properties of rhenium have been the subject of intense studies for about half of a XX century. The studies carried out in 1960s and 1970s focused on analysis of the electroreduction of perrhenates using polarographic techniques (Shropshire 1968; Letcher et al. 1970, 1971). Based on the results of experiments with a dropping mercury electrode, Shropshire (1968) suggested that in an acidic environment this process proceeds with formation of rhenium compounds with +VI oxidation state. Such generated Re(VI) undergoes a slow disproportionation reaction, which results in the formation of Re(IV) and Re(VII), according to Eq. (2.2).

$$3Re(VI) \leftrightarrows 2Re(VII) + Re(IV) \tag{2.2}$$

Letcher et al. (1970) reported an additional evidence that indicates that the reduction of perrhenate ion in 4M $HClO_4$ proceeds via an intermediate Re(V) state. Such behavior makes rhenium similar to technetium because also for the latter species with +V oxidation state (Tc(V)) are observed as the products of an electrode process in strongly acidic solutions (Chotkowski and Czerwiński 2012).

Till now, however, the products of the ReO_4^- electroreduction in an acidic environment have not been unequivocally identified. A general reduction scheme discussed by e.g., Horányi and Bakos (1993, 1994) include two main stages: formation of an oxide layer (1st stage) followed by generation of metallic rhenium (2nd stage). Both composition and structure of the oxide layer strongly depend on the experimental conditions, Eq. (2.3):

$$ReO_4^- \rightarrow Re \text{ (oxides layer)} \rightarrow Re \text{ (metallic)} \tag{2.3}$$

The above-described reactions scheme only partially reflects current state of knowledge on the reduction of pertechnetates.

Schrebler et al. (2001) analyzed the electroreduction of ReO_4^- on gold electrode in slightly acidic solutions (0.1 M Na_2SO_4, pH 2). At potentials lower than −0.75 V versus SCE, two parallel processes were observed: the formation of metallic rhenium and the hydrogen evolution. These researchers pointed out that generated hydrogen can facilitate reduction of perrhenate ions adsorbed on the gold surface. Such behavior differentiates rhenium from technetium since the electroreduction of TcO_4^- ions starts at potentials higher than the hydrogen evolution onset.

The formation of rhenium layers on a gold surface through the nucleation and growth mechanism during electroreduction of ReO_4^- in acidic solutions (0.1 M

Na_2SO_4; pH = 2) was also studied by Schrebler et al. (2001). They concluded that this process starts with a two-dimensional progressive nucleation (PN2D), followed by a 3D progressive nucleation under diffusional control ($PN3D_{dif}$) and, for a sufficiently long time, the process follows a progressive nucleation mechanism under charge transfer control ($PN3D_{CT}$). Unfortunately, there is lack of similar studies on nucleation and growth of metallic technetium. One may expect, however, that a similar mechanism should be applicable also for Tc.

A critical review on the electroreduction of perrhenate ions in acidic media was published in 2003 (Méndez et al. 2003). The authors concluded that the final products of this process strongly depend on the concentration of the acid in the solution. More recent works of Szabó and Bakos (2004) show that a nonsoluble ReO_2 is the main product of the perrhenates reduction in diluted H_2SO_4 (e.g. 0.1 M H_2SO_4). This aspect of the rhenium chemistry is somewhat similar to the chemical behavior of technetium because TcO_2, i.e., an analog to ReO_2, is considered as the main product of the TcO_4^- reduction. Szabó and Bakos reported that Re_2O_5 precipitate is formed as a non-soluble deposit in concentrated H_2SO_4 solutions (12 M H_2SO_4). On the other hand, electrogeneration of Tc_2O_5 has not been confirmed experimentally so far. Lawler et al. (2018) discussed the possibility of generation of so called "tech red," i.e., technetium species more volatile than Tc_2O_7. They assumed that the most probable identity of this mysterious form of technetium is Tc_2O_5 generated as a product of oxidation of technetium dioxide by oxygen in the presence of water at 250 °C. TcO^{3+}, which was described in detail for the first time by Poineau et al. (2013), is the form of Tc(V) observed in aqueous solutions at room temperature. Tc(V) may exist in various forms both in acidic and alkaline solutions and such behavior is different from what is observed for Mn and Re which species with + V oxidation state are stable only under very limited conditions. Thus, Mn(V) is observed only in concentrated alkaline solutions in contrast to Re(V), which can be generated in concentrated acidic solutions. These observations indicate periodicity of changes within the manganese group.

Some authors (e.g. Pihlar 1979) pointed out that technetium trioxide, TcO_3, can be generated on the electrodes during the pertechnetates reduction in acidic solutions. This behavior would make technetium similar to rhenium although TcO_3 is expected to be much less stable than ReO_3. The formation of rhenium(VI) oxide, ReO_3, as a result of the electrochemical reduction of perrhenates on a gold surface, has been confirmed by Schrebler et al. (2005) using the quartz microbalance technique. These authors found that electroreduction of the perrhenates in 0.1 M Na_2SO_4 + H_2SO_4 solution with pH of 2 at potentials before the onset of hydrogen evolution reaction (HER) leads to formation of not only rhenium (VI) oxide but also other rhenium oxide forms, i.e., ReO_2 and Re_2O_3. A scheme of these processes is shown in Fig. 2.9.

Szabó and Bakos suggested (2000) that in strongly acidic media rhenium(III) is produced as a result of rhenium dioxide disproportionation. Interestingly, respective disproportionation reactions are not observed for technetium dioxide. Although early works (e.g. Mazzocchin et al. 1974) suggested that Tc_2O_3 may play a role in reduction of technetium at lower potentials, there is lack of experimental evidence, which shows that Tc may be electroreduced exactly to Tc_2O_3.

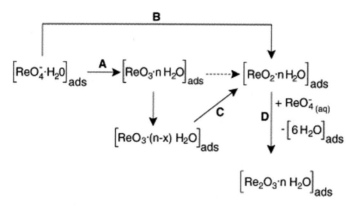

Fig. 2.9 Reaction formalism for the reduction of adsorbed ReO_4^- and the reduction of species formed onto electrode surface (EQCM-pc Au). In 0.1 M $Na_2SO_4 + H_2SO_4$ o pH $= 2$ at potential range: A (0.4–0.25 V), B (0.25 to −0.05 V), C (−0.05 to −0.24 V), D (−0.24 to −0.4 V) versus SCE (reprinted with permission from Schrebler et al. (2005). Copyright 2005 Elsevier)

The electroreduction of MnO_4^-, ReO_4^- and TcO_4^- ions strongly depends on the electrode material and, interestingly, the way how the type of the electrode material affects the electroreduction process depends on the manganese group element. CV curves recorded during electroreduction of the permanganates in acidic solution (0.5 M H_2SO_4) reveal a single, large and well-developed reduction peak, whose maximum shifts toward lower potentials when the electrode material is changed from platinum (1.27 V vs. SHE) through gold (1.22 V) to RVC (1.1 V). This peak is associated with the reduction of MnO_4^- to Mn^{2+}. An analysis of potentials of this peak allows concluding that the reaction in question occurs most easily on the surface of platinum electrodes for which the smallest overpotential was recorded.

An inspection of influence of the electrode material on recorded currents due to Tc reactions shows that the gold electrode provides the best conditions for analysis of electrochemical processes of Tc. This is caused by the fact that for this substrate, the technetium reactions generate best developed current signals, which are sufficiently separated from currents due to redox reactions of the electrode material or the electrolyte (Chotkowski and Czerwiński 2012). In contrast to the technetium case, the cyclic voltammetry curves recorded on a gold electrode in acidic solutions (e.g. 0.5 M H_2SO_4) containing perrhenates do not reveal well-developed current signals of ongoing electroreduction (Fig. 2.10). Current signals observed at potentials higher than 0.8 V are characteristic of gold surface oxidation and subsequent reduction. The reduction of pertechnetates is manifested by broad and poorly shaped wave, which indicates a multistep process. Oxidation of reduced Tc species can be analyzed on the basis of three main current peaks observed in anodic branch of the CV. Analysis of perrhenates (or pertechnetates) electroreduction using voltammetry curves recorded on a carbon electrode (e.g., reticulated vitreous carbon) is strongly complicated by the fact that for this electrode material only poorly developed reduction waves are observed for both elements.

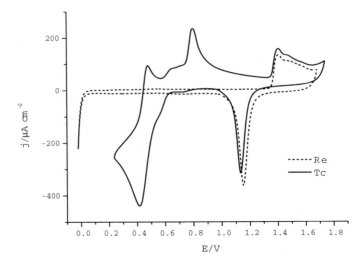

Fig. 2.10 Cyclic voltammetry curves recorded in 0.5 M H_2SO_4 in the presence of 1 mM ReO_4^- or TcO_4^- for gold electrode, scan rate = 50 mV s^{-1}

The shapes of cyclic voltammetric curves recorded with a platinum electrode are significantly different from those for the gold electrodes. The features similar for voltammetric curves recorded for Tc and Re include the decrease of hydrogen evolution currents, probably due to blocking effects, and existence of a strong, symmetric peak in anodic branch of the cyclic voltammetry curves. As compared with technetium, the electrochemical signals recorded for rhenium are shifted toward more negative values and this effect agrees with differences in standard redox potentials of both elements.

The first reduction peak of perrhenates is observed at potential of about 0.4 V (Fig. 2.11) and is associated with the reduction of ReO_4^- to ReO_2. Méndez et al. (2003) suggested that this process most likely includes reactions of preadsorbed hydrogen atoms, Eq. (2.3):

$$ReO_4^- + 3H_{ads} + H^+ \rightarrow ReO_2 + 2H_2O \qquad (2.3)$$

while reactions associated with charge transfer between the ReO_4^- anion and the electrode, Eq. (2.4), seem to be less important:

$$ReO_4^- + 4H^+ + 3e^- \rightarrow ReO_2 + 2H_2O \qquad (2.4)$$

This reaction scheme is supported by the fact that the perrhenates electroreduction occurs at lower potentials when electrodes made of nonadsorbing hydrogen materials are used.

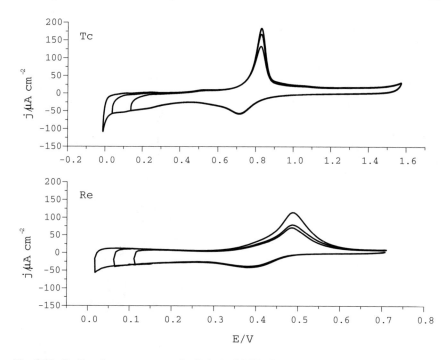

Fig. 2.11 Cyclic voltammetry curves for Pt in 1 mM KReO$_4$ or 1 mM KTcO$_4$ +0.5 M H$_2$SO$_4$ at 50 mV s^{-1} and various vertex potentials

The situation changes when cyclic voltammetry is preceded by chronoamperometric reduction of ReO$_4^-$. An analysis of CV curves recorded during the electrooxidation of such obtained rhenium layer reveals that the shape of obtained oxidation currents is similar, although not the same, to those observed for technetium. Application of gold as a working electrode in studies on electrochemical properties of rhenium deposits was reported by Zerbino et al. (2002) and by Szabó and Bakos (2004). Electrochemical analysis of rhenium layers by Zerbino et al. was carried out in solutions less acidic (1 M H$_2$SO$_4$) than those used by Szabó and Bakos who, in turn, prepared the rhenium deposits using chronoamperometry in solutions with acid concentrations of up to 12 M of H$_2$SO$_4$. Figure 2.12 presents a typical linear voltammogram (vs. RHE), which was recorded after long chronoamperomteric deposition of rhenium in 3–5 M H$_2$SO$_4$. Three anodic peaks that are observed for both rhenium and technetium represent oxidation of their forms with +IV, V and VI oxidation states. The peak centered at 0.7 V represents oxidation of ReO$_3$ while the one near 0.4 V is attributed to oxidation of ReO$_2$.

Another aspect of technetium chemistry important for thorough understanding of its electrochemical properties is adsorption of technetium compounds, especially pertechnetates, at electrode surfaces. These processes play an important role during electroreduction of MO$_4^-$ ions, especially for platinum surface. The adsorption of

Fig. 2.12 Potential sweep of the Au electrode covered with different rhenium species by polarization with 0.02 mA for 20 h in 3.0 M H_2SO_4 (1), 4.0 M H_2SO_4 (2) and finally in 5.0 M H_2SO_4 (3) containing 0.4 g L^{-1} Re_2O_7. Sweep rate: 2 mV s^{-1} (reprinted with permission from Szabó and Bakos (2004). Copyright 2004 Springer)

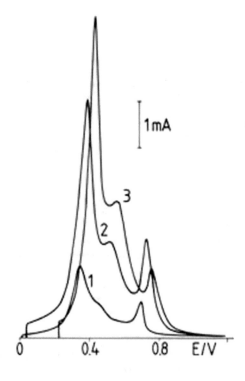

perrhenates on gold was investigated by Jusys and Bruckenstein (2000) using electrochemical quartz microbalance technique (Au-EQCM). These researchers found that in 0.1 M $NaReO_4$ solutions, the adsorption of ReO_4^- takes place at surface of a gold electrode at potentials of the double layer charging region (0.0–0.6 V SCE). Consequently, one may expect that the pertechnetates are adsorbed on gold electrodes also at potentials more positive than the onset of their electroreduction. Farrell et al. (1999) reported that pertechnetates can be effectively adsorbed on a polarized magnetite in 0.1 M NaCl solutions. As a result of this process, the technetium content in the solutions is reduced to a level lower than 900 pCi dm^{-3}. On the basis of radiometric measurements, Horányi and Bakos (1993) reported that removal of technetium species from the surface of platinized electrode starts in 1 M H_2SO_4 at potential of 0.75 V versus SHE.

Summary:

- Electrochemical properties of technetium are similar to those exhibited by its neighbors from the periodic table, especially rhenium. It should be noted, however, that electrochemical reactions of technetium generate numerous ionic species, both intermediates and final products, while for rhenium analogical processes involve mainly uncharged oxide molecules. Further on, oxides as ReO_2, Re_2O_5, ReO_3 are characteristic of rhenium, whereas for technetium TcO_2 is the only oxide well described in the literature.

- Potentials of technetium redox couples are shifted usually by 0.2–0.3 V toward more positive values as compared with the respective rhenium analogs. Not all redox systems known for Re are observed for Tc and vice versa.
- The tendency to dimerization or polymerization of Tc(IV) makes it similar to Mo(IV) and Ru(IV).
- Voltammetric curves of metallic technetium do not reveal the electrical double layer charging region, which makes technetium similar not only to rhenium but also to, e.g., ruthenium.
- Transport properties (e.g., diffusion coefficients, conductance) of technetium ionic forms are very similar to their rhenium analogs. They are practically the same for low concentrations of the ions ($\sim$mmol dm^{-3}).
- Influence of the electrode material on electroreduction of MO_4^- ions is observed for the manganese group elements (Mn, Tc, Re).
- Ionic forms of Re(VI) or Tc(VI) are unstable both in aqueous and nonaqueous solutions.
- Re(V) and Tc(V) forms are stabilized in concentrated acid solutions.

References

Ardon M, Pernick A (1974) The binuclear molybdenum(III)-aquo ion. Inorg Chem 13(9):2275–2277

Astheimer L, Schwochau K (1976) Electrochemical reduction of MnO_4^-, ReO_4^- and TcO_4^- in organic solvents. Preparation of tetraoxomanganate(VI), -technetate(VI) and -rhenate(VI). J Inorg Nucl Chem 38:1131–1134

Atlas of Eh-pH diagrams. Intercomparison of thermodynamic databases. (2005) Geological Survey of Japan Open File Report No. 419. National Institute of Advanced Industrial Science and Technology, pp 251–253

Bard AJ, Parsons R, Jordan J (eds) (1985) Standard potentials in aqueous solution. IUPAC, Marcel Dekker, pp 421, 450, 480

Baumann A, Yalçýntaţ E, Gaona X et al (2018) Thermodynamic description of Tc(IV) solubility and carbonate complexation in alkaline NaHCO$_3$–Na$_2$CO$_3$–NaCl systems. Dalton Trans 47:4377–4392

Boyd GE (1978) Osmotic and activity coefficients of aqueous HTcO$_4$ and HReO$_4$ solutions at 25 °C. Inor Chem 17(7):1808–1810

Cartledge GH (1971) Free energies of formation of hydrous oxides of technetium in its lower valencies. J Electrochem Soc 118(2):231–236

Chatterjee S, Hall GB, Johnson IE et al (2018) Surprising formation of quasi-stable Tc(VI) in high ionic strength alkaline media. Inorg Chem Front 5(9):2081–2091

Chotkowski M, Czerwiński A (2012) Electrochemical and spectroelectrochemical studies of pertechnetate electroreduction in acidic media. Electrochim Acta 76:165–173

Chotkowski M, Rogulski Z, Czerwiński A (2011) Spectroelectrochemical investigation of MnO$_2$ electro-generation and electro-reduction in acidic media. J Electroanal Chem 651:237–242

Cordoba de Torresi SI, Gorenstein A (1992) Electrochromic behavior of manganese dioxide films in slightly alkaline solutions. Electrochim Acta 37:2015–2019

Crompton TR, Crompton TPJ (2000) Battery reference book, 3rd ed, Chapters 6, 9, 10, 21, 25, 39. Newnes

Czerwiński A, Rogulski Z (2005) Patent RP–185542, Primary zinc-manganese cell (in polish)

de Zoubov, N, Pourbaix M (1966) Technetium chapter IV, section 11.2. In: Pourbaix M (ed) Atlas of electrochemical equilibria in aqueous solutions. Pergamon Press, p 298

Farrell J, Bostick WD, Jarabel RJ (1999) Electrosorption and reduction of pertechnetate by anodically polarized magnetite. Environ Sci Technol 33:1244–1249

Freeman DB, Mamantov G (1976) Linear sweep voltammetry of manganate(VII), manganate(VI), and manganate(V) in alkaline media. Electrochim Acta 21(4):257–261

Hamankiewicz B, Michalska M, Krajewski M et al (2014) Effect of electrode thickness on electrochemical performance of $LiMn_2O_4$ cathode synthesized by modified sol-gel method. Sol State Ion 262:9–13

Horányi G, Bakos I (1993) Combined radiometric and electrochemical study of the behaviour of Tc (VII) ions at gold and platinized surfaces in acidic media. J Appl Electrochem 23:547–552

Horányi G, Bakos I (1994) Investigation of the electrodeposition and behaviour of Re layers by coupled radiochemical and electrochemical methods. J Electroanal Chem 378(1–2):143–148

Icenhower JP, Qafoku NP, Zachara JM et al (2010) The biogeochemistry of technetium: a review of the behavior of an artificial element in the natural environment. Am J Sci 310:721–752

Ichimura A, Heineman WR, Vanderheyden JL et al (1984) Technetium electrochemistry. 2. Electrochemical and spectroelectrochemical studies of the bis(tertiary phosphine) (D) complexes trans-$[Tc^{III}D_2X_2]^+$ (X = chlorine or bromine). Inorg Chem 23(9):1272–1278

Isherwood D (1985) Application of the ruthenium and technetium thermodynamic data bases used in the EQ3/6 geochemical codes. UCRL-53594, University of California, Berkeley

Jusys Z, Bruckenstein S (2000) Electrochemical quartz crystal microbalance study of perchlorate and perrhenate anion adsorption on polycrystalline gold electrode. Electrochem Comm 2:412–416

Kirchhoff JR, Heineman WR, Deutsch E (1987) Technetium electrochemistry. 4. Electrochemical and spectroelectrochemical studies on the bis(tertiary phosphine or arsine (D))rhenium(III) complexes trans-$[ReD_2X_2]^+$ (X= Cl, Br). Comparison with the technetium(III) analogues. Inorg Chem 26:3108–3113

Kirchhoff JR, Heineman WR, Deutsch E (1988) Technetium electrochemistry. 6. Electrochemical behavior of cationic rhenium and technetium complexes in aqueous and aqueous micellar solutions. Inorg Chem 27:3608–3614

Lawler KV, Childs BC, Czerwinski KR et al (2018) Unraveling the mystery of "tech red"—a volatile technetium oxide. Chem Commun 54:1261–1264

Lee J, Maskell WC, Tye F (1977) The electrochemical reduction of manganese dioxide in acidic solutions. Part 1. Voltammetric peak 1. J Electroanal Chem 79(1):79–104

Lee J, Maskell WC, Tye F (1980) The electrochemical reduction of manganese dioxide in acidic solutions Part III. Voltammetric peak 3. J Electroanal Chem 110(1–3):145–158

Letcher DW, Cardwell TJ, Magee RJ (1970) A study of the electrochemical behaviour of perrhenate ion in aqueous solution. Electroanal Chem 25:473–479

Letcher DW, Cardwell TJ, Magee RJ (1971) The polarographic reduction of perrhenate ion. Part II. Electroanal Chem 30:93–99

Mausolf E, Poineau F, Droessler J et al (2011) Spectroscopic and structural characterization of reduced technetium species in acetate media. J Radioanal Nucl Chem 288(3):723–728

Mazzocchin GA, Magno F, Mazzio U et al (1974) Voltammetric behaviour of aqueous technetate(VII) ion. Inorg Chim Acta 9:263–268

Méndez E, Cerdá MF, Castro Luna AM et al (2003) Electrochemical behavior of aqueous acid perrhenate-containing solutions on noble metals: critical review and new experimental evidence J Coll Interfac Sci 263:119–132

Norwell VE, Mamantov G (1977) Optically transparent vitreous carbon electrode. Anal Chem 49(9):1470–1472

Paffett MT, Anson FC (1983) Electrochemistry of the trinuclear aquo MoIV3 and MoIII3 ions in acidic media. Inorg Chem 22:1347–1355

Pihlar B (1979) Electrochemical behavior of technetium(VII) in acidic medium. J Electroanal Chem 102:351–365

Poineau F, Weck PF, Burton-Pye BP et al (2013) Reactivity of HTcO$_4$ with methanol in sulfuric acid: Tc-sulfate complexes revealed by XAFS spectroscopy and first principles calculations. Dalton Trans 42(13):4348–4352

Poineau F, Burton-Pye BP, Sattelberger AP (2018) Speciation and reactivity of heptavalent technetium in strong acids. New J Chem 42:7522–7528

Povar I, Spinu O (2016) Ruthenium redox equilibria 3. Pourbaix diagrams for the systems Ru-H$_2$O and Ru-Cl-H$_2$O. Electrochem Sci Eng 6(1):145–153

Rard JA (1985) Chemistry and thermodynamics of ruthenium and some of its inorganic compounds and aqueous species. Chem Rev 85(1):1–39

Rard JA, Rand MH, Anderegg G et al (1999) Chemical thermodynamics. of technetium, vol 3. Elsevier

Rogulski Z, Chotkowski M, Czerwiński A (2006) Electrochemical behavior of MnO$_2$/RVC system. J New Mat Electrochem Sys 9(4):401–408

Rouschias G (1974) Recent advances in the chemistry of rhenium. Chem Rev 74(5):531–566

Schrebler R, Cury P, Orellana M et al (2001) Electrochemical and nanoelectrogravimetric studies of the nucleation and growth mechanisms of rhenium on polycrystalline gold electrode. Electrochim Acta 46:4309–4318

Schrebler R, Cury P, Suárez E et al (2005) Study of the electrodeposition of rhenium thin films by electrochemical quartz microbalance and X-ray photoelectron spectroscopy. Thin Solid Films 483(1–2):50–59

Shropshire JA (1968) Electrochemical reduction of perrhenate in sulfuric acid solutionJ. Electroanal Chem 16(2):275–278

Szabó S, Bakos I (2000) Electroreduction of rhenium from sulfuric acid solutions of perrhenic acid. J Electroanal Chem 492:103–111

Szabó S, Bakos I (2004) Electrodeposition of rhenium species onto a gold surface in sulfuric acid media. J Solid State Electrochem 8:190–194

Tikkanen A (2008) Encyclopedia of Brittanica. Noble metals. https://www.britannica.com/science/noble-metal. Accessed 7 Aug 2019

Tisato F, Refosco F, Moresco A et al (1990) Synthesis and characterization of technetium(v) and rhenium(v) oxo-complexes with Schiff-base ligands containing the ONN donor-atom set. Molecular structure of trans-dichloro-oxo[1-(8′-quinolylimino methyl)-2-naphtholato-NN′O]technetium(v). J Chem Soc Dalton Trans 2225–2232

Vongsouthi N (2009) Spéciation du technétium-99 en milieu acide noncomplexant: effet Eh-pH, Ph.D. thesis, Université de Nantes, UFr Sciences et Techniques

Warwick P, Aldridge S, Evans N et al (2007) The solubility of technetium(IV) at high pH. Radiochim Acta 95:709–716

Zerbino JO, Castro Luna AM, Zinola CF et al (2002) Comparative study of electrochemical and optical properties of rhenium deposited on gold and platinum. J Braz Chem Soc 13(4):510–515

Chapter 3
Technetium Coordinated by Inorganic Ligands in Aqueous and Nonaqueous Solutions

Among the several oxidation states of technetium, which may be observed in organic and inorganic media, the most stable and, consequently, the most important are +4 (as TcO_2) and +7 (as TcO_4^-). The transport properties of the latter were investigated by Boyd (1978), Rard and Miller (1991) and Könnecke et al. (1997).

Table 3.1 presents the activity coefficients obtained for $NaTcO_4$ and $HTcO_4$. The data are taken from Rard and Miller (1991), who reevaluated earlier results published by Boyd (1978). The activity coefficients for the highest acid concentration were calculated using a graphical extrapolation from lower to higher concentrations region. These authors assumed the following equation for osmotic coefficient, ϕ, where $A^* = 0.34733$ ($=A/(1.5)^3$) $kg^{1/2} \cdot mol^{-1/2}$ at 25 °C is given by Eq. (3.1):

$$\phi = 1 - \frac{A^*|Z_+ Z_-|}{I_m}$$
$$\times \left[\left(1 + 1.5 \cdot \sqrt{(I_m)}\right) - 2\log_e\left(1 + 1.5 \cdot \sqrt{(I_m)}\right) - \left(1 + 1.5 \cdot \sqrt{(I_m)}\right)^{-1}\right]$$
$$+ \sum_{i=4}^{p} A_{i-3} \cdot m^{i/4} \tag{3.1}$$

and the activity coefficient, $\gamma_\pm$, where m stands for molality while A' is equal to 1.1722 $mol^{-1/2} \cdot kg^{1/2}$ at 25 °C, can be given by Eq. (3.2) (Rard et al. 1999):

$$\log_e \gamma_\pm = -\frac{A'|Z_+ Z_-|\sqrt{I_m}}{1 + 1.5 \cdot \sqrt{I_m}} + \sum_{i=4}^{p} \frac{(i/4 + 1)}{i/4} \cdot A_{i-3} \cdot m^{i/4} \tag{3.2}$$

They also concluded that the above equations give the most accurate fit to the experimental data when a least-square fitting is applied.

The results of the analysis taken by Rard turned out to be in good agreement with data published later by Könnecke et al. (1997) who analyzed $HTcO_4$ and $NaTcO_4$ − NaCl systems. They applied Pitzer's approach in calculation of parameters describing the interaction of ions (Pitzer 1991). The osmotic coefficient was calculated according to Eq. (3.3). a_w is water activity, M_w is equal to 18.0152 $g \cdot mol^{-1}$ and m_i is the molality

© Springer Nature Switzerland AG 2021
M. Chotkowski and A. Czerwiński, *Electrochemistry of Technetium*,
Monographs in Electrochemistry, https://doi.org/10.1007/978-3-030-62863-5_3

Table 3.1 Mean molar activity coefficients and water activity for NaTcO₄ and HTcO₄ solutions at 25 °C (Rard and Miller 1991)

Molality/mol·kg^{-1}	$\gamma_\pm$	a_{H_2O}	$\gamma_\pm$	a_{H_2O}
	NaTcO₄		HTcO₄	
0.1	0.7455	0.99669	0.7681	0.99666
0.2	0.6937	0.99348	0.7239	0.99338
0.3	0.6605	0.99034	0.7009	0.99009
0.4	0.6350	0.98726	0.6871	0.98677
0.5	0.6141	0.98424	0.6785	0.98342
0.6	0.5963	0.98127	0.6733	0.98004
0.7	0.5808	0.97835	0.6704	0.97661
0.8	0.5672	0.97545	0.6694	0.97314
0.9	0.5552	0.97258	0.6699	0.96961
1.0	0.5444	0.96973	0.6716	0.96604
1.5	0.5050	0.95560	0.6954	0.94726
2.0	0.4809	0.94141	0.7408	0.92661
2.5	0.4656	0.92700	0.8080	0.90365
3.0	0.4559	0.91234	0.9004	0.8780
3.5	0.4499	0.8974	1.024	0.8495
4.0	0.4466	0.8823	1.186	0.8178
4.5	0.4453	0.8669	1.399	0.7832
5.0	0.4458	0.8513	1.678	0.7456
5.5	0.4477	0.8354		
6.0	0.4509	0.8193		
6.5	0.4551	0.8030		
7.0	0.4601	0.7865		
7.5	0.4655	0.7701		
8.0	0.4708	0.7540		

of ion i (in mol·kg^{-1} H₂O).

$$\phi = -\frac{1000\ln(a_w)}{M_w \Sigma m_i} \tag{3.3}$$

The appropriate binary and mixing Pitzer parameters for $Na^+/Cs^+/TcO_4^-/H_2O$ mixtures used by Könnecke et al. (1997) are listed in Table 3.2.

The electrical conductance of TcO_4^- (25 °C), Λ^0, is equal to 55.4 ± 0.5 S·cm² Eq^{-1} (Rard et al. 1999). The conductance of potassium pertechentate solutions at 25 °C was studied in early works of Schwochau and Astheimer (1962). An analysis of their results performed by Rard et al. (1999) indicates that the following Debye–Hückel–Onsager limiting law, where I_c is the ionic strength in molar units, can be

Table 3.2 Pitzer parameters in $Na^+/Cs^+/TcO_4^-/H_2O$ system at 25 °C (Könnecke et al. 1997)

Binary parameters			
i/j	$\beta^{(0)}$	$\beta^{(1)}$	C^ϕ
Na^+/TcO_4^-	0.01111	0.1595	0.00236
Cs^+/TcO_4^-	−0.1884	−0.1588	0
Mixing parameters			
i/j/k	ϕ_{ijk}	ψ_{ijk}	
$TcO_4^-/Cl^-/Na^+$	0.067	−0.0085	
$TcO_4^-/Cl^-/Cs^+$	0.067	−0.0011	

applied here, Eq. (3.4):

$$\Lambda = \Lambda^0 - \left(0.2300 \cdot \Lambda^0 + 60.64\right) \cdot I_c^{1/2} \tag{3.4}$$

The recommended value of the tracer diffusion coefficient of the pertechnetates in water at 25 °C is equal to $(1.48 \pm 0.01) \cdot 10^{-5}$ cm$^2 \cdot$s^{-1}. According to Astheimer and Schwochau (1976), the latter parameter decreases in electrolyte solutions and in 1 mM solutions of TcO_4^- in 1 M NaOH and 1 M LiCl has the values of $(1.27 \pm 0.01) \cdot 10^{-5}$ and $(1.24 \pm 0.01) \cdot 10^{-5}$ cm$^2 \cdot$s^{-1}, respectively.

Tc(IV) may be contained in various ionic and nonionic species and this leads to significant discrepancies between the standard potential values reported by various authors for the $TcO_4^-/TcO_2 x H_2O$ system (Rard et al. 1999). The net reaction of the pertechnetates reduction to the technetium dioxide is presented by Eq. (3.5):

$$TcO_4^- + 4H^+ + 3e^- \leftrightarrows TcO_2 \cdot xH_2O_{(s)} + (2 - x)H_2O \tag{3.5}$$

Unfortunately, the stoichiometry of the hydrated technetium dioxide which is generated in this reaction is still unclear. An arbitrary value of x equal to 1.6 was assumed in the review by Rard et al. (1999), but it is known that the degree of the TcO_2 hydration changes very slowly with time in aqueous solutions. Yalçıntaş studied Tc(IV) speciation in the NaCl environment assuming x equal to 0.6 (e.g. Yalçintas et al. 2016). Thus, it is possible that under equilibrium conditions x may reach a value equal to 0 (see: Grambow et al. 2020).

The experimentally determined value of the standard redox potential, $E^\ominus$, of the system given by Eq. (3.1) with TcO_2 (am, hyd, fresh) as the reaction product is equal $E^\ominus = 0.746 \pm 0.012$ V at 25 °C (Rard et al. 1999). On the other hand, the standard potential for TcO_4^-/TcO_2 redox system calculated on the basis of solubility of TcO_2 (am, hyd, aged) is slightly higher than the above value and is equal to $E^\ominus = 0.757 \pm 0.006$ V (see: Grambow et al. 2020). The experimental emf of this system determined in low ionic strength solutions ($I_m \leq 10$ mmol·kg^{-1}) at 25 °C is represented by Eq. (3.6) (E, in V):

$$E = E^0 - 0.0789 \cdot pH + 0.0197 \cdot \log_{10}\left(a_{TcO_4^-}\right) \tag{3.6}$$

Poorly bounded hydration water has little effect on the TcO_2. For this reason, one may write Eq. (3.7):

$$TcO_2(am, hyd) + H_2O \leftrightarrows TcO(OH)_2(aq) \tag{3.7}$$

The standard redox potential of $TcO_4^-/TcO(OH)_2(aq)$ couple calculated from Eq. (3.8):

$$TcO_4^- + 4H^+ + 3e^- \leftrightarrows TcO(OH)_2(aq) + H_2O \tag{3.8}$$

turned out to be definitely lower than the one given for the reaction (3.5) and is equal to $E^\ominus = 0.595 \pm 0.046$ V at 25 °C (Grambow et al. 2020).

In an acidic medium, the technetium (IV) oxohydroxide can undergo a dimerization process, which leads to the formation of $Tc_2O_2(OH)_2^{2+}$. The standard redox potential of the $TcO_4^-/Tc_2O_2(OH)_2^{2+}$ couple was calculated using thermodynamic data for reaction (3.9):

$$2TcO_4^- + 10H^+ + 6e^- \leftrightarrows Tc_2O_2(OH)_2^{2+} + 4H_2O \tag{3.9}$$

and is equal to $E^\ominus = 0.731 \pm 0.011$ V (see: Grambow et al. 2020).

It is worth mentioning that apart from the Tc(IV) forms mentioned above also other Tc(IV) containing species may percolate in the equilibrium of technetium compounds (e.g., Rard et al. 1999). Carbonyl complexes of technetium with halides, which play an important role in nuclear medicine, are a good example here. In Table 3.3 are given the equilibrium constants for selected technetium species.

Table 3.3 Experimental equilibrium data for selected technetium systems

reaction	$\log_{10}K$	conditions	reference
$TcO_2 \cdot 0.6H_2O + \frac{2}{3}H^+ \leftrightarrows \frac{1}{3}Tc_3O_5^{2+} + 0.93H_2$	-1.53 ± 0.15		
$TcO_2 \cdot 0.6H_2O + 0.4H_2O \leftrightarrows TcO(OH)_2$	-8.80 ± 0.50	$I=0, T=25$ °C	(Yalçıntaş 2016)
$TcO_2 \cdot 0.6H_2O + 1.4H_2O \leftrightarrows TcO(OH)_3^- + H^+$	-19.27 ± 0.06		
$TcO(OH)_2(aq) + 2H^+ \leftrightarrows TcO^{2+} + 2H_2O$	<4		
$TcO(OH)_2(aq) + H^+ \leftrightarrows TcO(OH)^+ + H_2O$	2.500 ± 0.300		
$TcO(OH)_2(aq) + H_2O \leftrightarrows TcO(OH)_3^- + H^+$	-10.9 ± 0.4	$I=0, T=25$ °C	(See: Rard 1999, Guillaumont 2003)
$TcO_2 \cdot 1.6H_2O(s) \leftrightarrows TcO(OH)_2(aq) + 0.6H_2O$	-8.400 ± 0.500		
$TcO(OH)_2(aq) + CO_2(g) \leftrightarrows TcCO_3(OH)_2(aq)$	1.100 ± 0.300		
$TcO(OH)_2(aq) + CO_2(g) + H_2O \leftrightarrows TcCO_3(OH)_3^-$	-7.200 ± 0.600		
$2TcO(OH)_2(aq) + 2H^+ \leftrightarrows Tc_2O_2(OH)_2^{2+} + 2H_2O$	-12.99 ± 0.41	$I=0, T=25$ °C	(See: Grambow 2020)
$Tc(CO)_3^+ + X^- \leftrightarrows Tc(CO)_3X$	Cl⁻: 1.23; Br⁻: 1.26; I⁻: 2.67; CNS⁻: 697		
$Tc(CO)_3X + X^- \leftrightarrows Tc(CO)_3X_2^-$	Cl⁻: 0.17; Br⁻: 0.21; I⁻: 1.26; CNS⁻: 106	* $I=4$ M (NaX + NaClO₄), $T=22\pm2$ °C,	(Gorshkov 2003)
$Tc(CO)_3X_2^- + X^- \leftrightarrows Tc(CO)_3X_3^{2-}$	Cl⁻: 0.05; Br⁻: 0.07; I⁻: 0.66; CNS⁻: 13		
$Tc(OH_2) + Cl^- \leftrightarrows TcCl_6^{2-}$	-0.0969	4M HCl, $T=25$ °C	(Rajec 1981)

* Taking into account presence of the water molecules in the respective equations the values determined in solutions with $I=4$ should be multiplied by 51. Uncertainities at a level of 20 %

3.1 Acidic solutions

The electrochemical reactions of ionic forms of technetium in acidic solutions have been of interest to scientists for many years. Electroreduction of the pertechnetates in acidic solutions is usually considered as a process which leads to formation of Tc(IV) or Tc(III) as final products.

Kuzina and coworkers reported in 1962 that the reduction of the pertechnetates in acidic solutions is a three electron process with formation of TcO_2 (Kuzina et al. 1962). Later works of Salaria et al. (1963a, b), Grassi et al. (1978) or Russell and Cash (1978) found that an irreversible transfer of three electrons is the first step in the Tc(VII) electroreduction. The current due to this reaction is poorly separated from the signal due to a subsequent reaction of Tc(IV) → Tc(III) as reported by Salaria et al. (1963a). These authors also concluded that the reduction of Tc(VII) in noncomplexing media at pH higher than 4 leads to the formation of Tc(IV). The αn value calculated by Grassi and coworkers for the first anodic wave related to oxidation of diffusible Tc(III) was found to be equal to ca. 0.5.

Some papers point out a possible generation of unstable Tc(V) or Tc(VI) intermediates during electroreduction of the pertechnetates. Pihlar (Pihlar 1979) dealt with the problem of TcO_4^- electroreduction on a mercury electrode in slightly acidified solutions of NaCl with pH from 1.84 to 4.07 and with or without addition of Triton X-100 as a detergent (maximum concentration of 0.002%). He observed that a decrease in H^+ concentration at the electrode surface slows down the pertechnetates reduction rate and promotes formation of oxides containing technetium with a lower oxidation state that are adsorbed at the electrode surface. The kinetic parameters of the reduction reactions and the values of the heterogeneous rate constant $k_{f,h}^0$ were calculated on the basis of obtained polarograms. The results of the analysis are summarized in Table 3.4. $\log k_{f,h}^0$ is equal to $(1.806 \div 1.743) \cdot pH$ while $E_{1/2}$ is calculated as $(0.142 \div 0.148) \cdot pH$.

Pihlar (1979) analyzed three different mechanisms of the pertechnetates electroreduction assuming only the electrochemical steps involved in this process. He calculated several electrochemical parameters, including $\alpha n = 0.6 \pm 0.1$, $\left(\frac{\partial \log i}{\partial pH}\right)_{E,c} = -2$; $\left(\frac{\partial \log i}{\partial \log c}\right)_{E,pH} = 1$. It was also concluded that the most accurate description of the

Table 3.4 Electrochemical kinetic parameters for the reduction of $5.0 \cdot 10^{-5}$ M TcO_4^- at dropping mercury electrode, $t = 2.00$ s, $m = 1.87$ mg·s^{-1} (Pihlar 1979)

pH	Parameter	
	$E_{1/2}$ versus SCE	$\log k_{f,h}^0$
1.84	−0.133	−1.42
2.16	−0.178	−1.96
2.47	−0.218	−2.445
2.82	−0.272	−3.071
3.08	−0.318	−3.61

overall process includes a set of consecutive reactions (3.10)–(3.12) with the initial TcO_3 formation considered as the rate-determining step:

$$TcO_4^- + 2H^+ + e^- \overset{rds}{\rightarrow} TcO_3 + H_2O \tag{3.10}$$

$$TcO_3 + e^- \leftrightarrows TcO_3^- \tag{3.11}$$

$$TcO_3^- + 3H^+ + 2e^- \leftrightarrows TcO(OH) + H_2O \tag{3.12}$$

Cobble and coworkers reported the estimated value of the standard potential of TcO_4^-/TcO_3 redox couple in aqueous solutions equal to 0.7 V (Cobble et al. 1953). They noted that TcO_3 is unstable in water and decomposes to pertechnetic acid and technetium dioxide.

It is well established that Tc(VI) and Tc(V) are unstable in aqueous acidic solutions and undergo disproportionation. For instance, Courson et al. (1999) suggested the following reactions of the Tc(IV) and Tc(V) disintegration (3.13)–(3.14):

$$2TcO_4^{2-} \rightarrow TcO_4^{3-} + TcO_4^- \tag{3.13}$$

$$2TcO_4^{3-} \overset{H^+}{\rightarrow} TcO_4^{2-} + TcO^{2+} \tag{3.14}$$

TcO^{2+} ions may undergo a hydrolysis with the formation of TcO_2 (see: Table 3.3). Recent EXAFS studies have shown that Tc(V) forms generated in a strongly acidic environment (13 M H_2SO_4) have TcO^{3+} core (Poineau et al. 2013). Lawson and coworkers (Lawson et al. 1984) reported that at pH 1-2 pertechnetates are reduced to Tc(IV) species adsorbed at the surface of a carbon electrode. They pointed out that also some Tc(V) forms are adsorbed at the electrode surface in formate solutions with pH of 3.

Chotkowski and Czerwiński (2012) investigated the electrochemical properties of technetium species in aqueous solutions with a wide range of concentration of sulfuric acid. Figure 3.1 presents typical CVs recorded in solutions containing from 1 to 4 M of H_2SO_4 with addition of 0.5 mM TcO_4^-. The voltammetry curves reveal currents due to redox processes of technetium at potentials lower than ca. 1.1 V versus SHE as well as currents due to oxidation and reduction of gold substrate surface which are located above ca. 1 V. An analysis shows that an increase in the solution acidity enhances currents of the reduction peak 1, which is attributed to the generation of Tc(VI) and Tc(V) on the electrode surface (Fig. 3.1). The next reduction signal, a very wide peak at ca. 0.5 V, is associated with formation of Tc(III) and Tc(IV) species. Anodic section of the voltammetric curves recorded after changing direction of the potential scan reveals three partially overlapping current peaks located below ca.

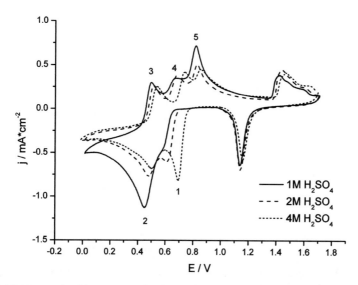

Fig. 3.1 Cyclic voltammograms recorded in electrolytes containing 1.19 mM KTcO$_4$ and various concentrations of H$_2$SO$_4$, $v = 0.05$ V s^{-1}, $E_{start} = 0.9$ V (reprinted with permission from Chotkowski and Czerwiński (2012) Copyright 2012 Elsevier)

1.1 V and related to electrochemical reactions of technetium. The first of them, peak 3 is attributed to oxidation of Tc(III) forms to Tc(IV). The smallest of the anodic signals, peak 4, is related to the oxidation of Tc(V) and Tc(VI) to Tc(VII). The last of the anodic signals, asymmetric peak 5, is connected with oxidation of polymeric Tc(IV) forms to Tc(VII) and its currents have the highest values in 1 M H$_2$SO$_4$ solution.

A decrease in the scan rate and acid concentration leads to disappearance of peaks 3 and 4 (Fig. 3.1) (e.g. Horányi and Bakos 1994). On the basis of the results of radioelectrochemical methods, these authors concluded that the reduction of the pertechnetates in 1 M H$_2$SO$_4$ at potentials below ca. 0.5 V versus RHE leads to simultaneous formation of soluble Tc(IV) forms as well as a layer of Tc species adsorbed at the gold electrode surface (Fig. 3.2).

These authors also suggest possible generation of Tc(VI) and Tc(V) species in the initial steps of the pertechnetates electroreduction.

An analysis of results of rotating disc-ring electrode experiments performed by Chotkowski and Czerwinski (2012) indicates the formation of Tc(V) species in sulfuric acid solutions. Due to their instability under experimental condition, these species could not be detected using UV–Vis spectroscopy. Tc(VI) are probably generated in this system as follows from an analysis of the standard entropy of formation of TcO$_3$. The experimentally determined standard entropy of the reaction was equal to -53.1 J·M^{-1}·K^{-1}, which is close to the value calculated on the basis of the literature data (Rard et al. 1999).

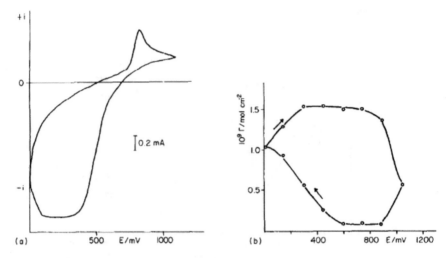

Fig. 3.2 a Voltammetric curve recorded in 1 M H_2SO_4 in the presence of TcO_4^- ions with initial concentration of 6.3×10^{-4} mol·dm^{-3} after 50 cycles at sweep rate of 2.5 mVs^{-1}; **b** corresponding voltradiometric curve (Reprinted with permission from Horányi and Bakos (1994) Copyright 1994 Elsevier)

Further experiments carried out by Chotkowski et al. (2018a), which were focused on the electroreduction of the pertechnetates in concentrated sulfuric acid solutions confirmed the generation of stable technetium species at intermediate oxidation states. The currents associated with the first stage of the TcO_4^- reduction are shifted toward higher potential values when the acidity increases: from 0.4 V in 0.5 M H_2SO_4 to 0.75 V in 10 M H_2SO_4 (versus Hg, Hg_2SO_4, 0.5 M H_2SO_4). An additional chronopotentiometric measurements reveal Tc(V) as a product of the Tc(VII) reduction ($n_e = 2$). These results are in line with those reported by Poineau (Poineau et al. 2013), which show the existence of stable Tc(V) forms in strongly acidic solutions.

The chronovoltamperometric measurements carried out in 4 M H_2SO_4 (Chotkowski and Czerwinski 2012) were analyzed in terms of typical reaction models, which involve mixed electrochemical–chemical mechanisms (see: Compton and Banks 2011), Eqs. (3.15)–(3.20):

ECE:

$$A + e^- \leftrightarrows B; B \leftrightarrow C(rds); C + e^- \leftrightarrows D \qquad (3.15)$$

$$\partial Ep/\partial \log v = 30; \partial Ep/\partial c = 0 \qquad (3.16)$$

DISP1:

$$A + e^- \leftrightarrows B; B \leftrightarrow C(rds); B + C \leftrightarrows A + D \qquad (3.17)$$

$$\partial Ep/\partial \log v = 30; \; \partial Ep/\partial c = 0 \qquad (3.18)$$

DISP2:

$$A + e^- \leftrightharpoons B; \; B \leftrightarrow C; \; B + C \leftrightharpoons A + D(rds) \qquad (3.19)$$

$$\partial Ep/\partial \log v = 20; \; \partial Ep/\partial c = 20 \qquad (3.20)$$

where *rds* is the rate-determining step.

The measurements allowed determining $\partial Ep/\partial \log v$ and $\partial Ep/\partial c$ values for the pertechnetates reduction in the acidic solutions. These values were equal to 14 and 0, respectively, clearly indicating that the overall process is significantly more complex than simple ECE or DISP mechanisms.

Chotkowski and Czerwiński (2014) point to a significant influence of the electrode material on the process of the pertechnetates electroreduction in acidic solutions (Fig. 3.3). Only gold electrodes reveal well-shaped current peaks due to Tc redox

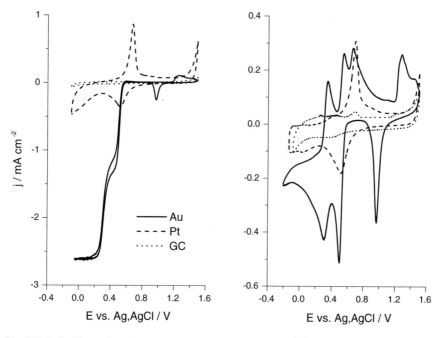

Fig. 3.3 Left side: cyclic voltammograms of gold, platinum and glassy carbon rotating disk electrodes. Rotation speed of 1600 rpm; 4 M H_2SO_4 + 0.55 mM $KTcO_4$, $v = 33.3$ mV·s^{-1}. Right side: cyclic voltammogram of gold, platinum and glassy carbon electrodes in 4 M H_2SO_4 + 1 mM $KTcO_4$, stationary conditions, $v = 50$ mV·s^{-1} (reprinted with permission from Chotkowski and Czerwiński (2014b) Copyright 2014 Elsevier)

reactions, i.e., two very strong reduction waves and three oxidation signals. The electrochemical signals of TcO_4^- reaction on a platinum electrode are very similar to those reported for electrolytes containing rhenium (Méndez et al. 2003; Szabó and Bakos 2004). The latter reveals a strong adsorption of ReO_4^- ions on the electrode surface and formation of ReO_2 oxides as well as Re_2O_5 and ReO_3 during the perrhenates electroreduction. Currents due to oxidation and reduction of Tc on glassy carbon (GC), which was the third type of the electrode material studied by Chotkowski and Czerwiński (2014), are much less developed as compared to Au and Pt substrates (Fig. 3.3).

Results of the electrochemical measurements performed on Au, Pt and GC (glassy carbon) electrodes under hydrodynamic conditions led to the conclusion that well-developed limiting currents of redox reactions of technetium are observed only for the gold surface (Fig. 3.4). These currents are due to reduction of Tc(VII) ions to Tc(III, IV). Activation energy of these reactions is equal to a dozen of kJ mol^{-1} and decreases with an increase in the acid concentration (Fig. 3.4).

Several papers focus on studies on the equilibrium between Tc(IV) and Tc(III) forms. Grassi et al. (1979), for example, investigated the oxidation of 0.3 mM Tc(III) in slightly acidic media containing Na_2SO_4 and $NaHSO_4$ (pH from 1.6 to 3.25). The Tc(III) was obtained by a coulometric reduction of the pertechnetates at -0.5 V versus SCE in a solution with pH 1.5 and containing 0.5 M $NaHSO_4 - Na_2SO_4$ and 0.6 mM TcO_4^-. Grassi and coworkers proposed the reaction (3.21) as the one explaining chemical behavior of the observed quasi-reversible Tc(III)/Tc(IV)

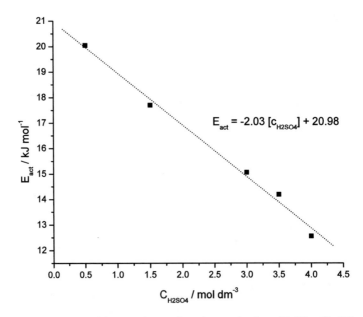

Fig. 3.4 Activation energy of the pertechnetate ions electroreduction to Tc(III and/or IV) species as a function of the sulfuric acid concentration calculated on the basis of data from hydrodynamic (Au-RDE) experiments (reprinted with permission from Chotkowski and Czerwiński (2014b) Copyright 2014 Elsevier)

process:

$$TcO_2 + 2H^+ + e^- \leftrightarrows TcO^+ + H_2O \qquad (3.21)$$

The equilibrium red–ox potential that corresponds to this reaction is given by Eq. (3.22):

$$E = 0.319 - 0.1182 \cdot pH - 0.0591 \cdot \log\left[TcO^+\right] \qquad (3.22)$$

These authors also stated that the oxidation of Tc(III) to Tc(IV) is faster than subsequent oxidation of the Tc(IV) to the pertechnetates. Taking into account, the hydrolysis constant of Tc(IV) (Table 3.3) one may expect the existence of more than one hydrolyzed form of Tc(IV) in the aqueous solutions. It should be stressed that one electron Nernstian slope predicted by Eq. (3.21) is expected also for Tc^{3+}, which is suggested by other authors as the Tc(IV) reduction product. The cationic structure of the Tc(III) species is in line with the results reported by other researchers (e.g. Salaria et al. 1963a; Rulfs et al. 1967). Oxidation of the Tc(III) may lead to formation of TcO^{2+}. Salaria with coworkers proposed Tc^{3+} as the product of polarographic electroreduction of the pertechnetates at pH < 4, Eq. (3.23):

$$TcO_4^- + 8H^+ + 4e^- \leftrightarrows Tc^{3+} + 4H_2O \qquad (3.23)$$

These authors assumed this process as an irreversible one with the half-wave potential given by Eq. (3.24):

$$E_{1/2} = 0.12 - 0.11_3 \cdot pH \ (0.25M \ Na_2SO_4 + H_2SO_4) \qquad (3.24)$$

Grassi reported that the oxidation of Tc(III) to Tc(IV) is accompanied by generation of some electro-inactive Tc(IV) species. Spitsyn proposed its structure as a $[Tc(OH)_2(SO_4)_2]^{2-}$ complex. Today it is known that the reduced Tc(III/IV) species can exist in aqueous solutions also as dimeric (or polymeric?) forms with $[Tc(\mu-O)_2Tc]^{4+ \, or \, 3+}$ core structure (Mausolf et al. 2011). Recent research carried out in strongly acidic solutions indicates that reduced Tc species can also exist as polymers with $[Tc(\mu-O)Tc]^{6+}$ core structure (Poineau et al. 2018).

Chotkowski and Czerwiński have shown that Tc(IV) polymeric species are electrooxidized at potentials higher than 0.7 V (Chotkowski and Czerwinski 2014a, b). These authors investigated electroreduction of the pertechnetates and subsequent oxidation of such obtained reduced technetium species in strongly acidic media using UV–Vis spectroelectrochemistry, especially voltabsorbommetry (VAB).

The simultaneous use of both spectroscopic and electrochemical methods in examination of organic or inorganic compounds has been applied by many authors (see e.g., Heineman et al. 1984; Kim et al. 1995; Kulesza et al. 1998; Astuti et al. 2004; Duluard et al. 2010; Wang and He 2012). The spectroelectrochemical techniques are particularly useful for determining the mechanism of electrochemical processes with participation of species dissolved in solution. The VABs are calculated as a time

Fig. 3.5 Cyclic voltammograms and voltabsorbommograms (VABs) for wavelengths 244, 320, 440 and 504 nm of a OTTL-RVC electrode in 4 M H_2SO_4 + 0.5 mM $KTcO_4$, $v = 1\ mV\cdot s^{-1}$, $E_{start} = 0.6$ V (reprinted with permission from Chotkowski and Czerwiński (2014a) Copyright 2014 Elsevier)

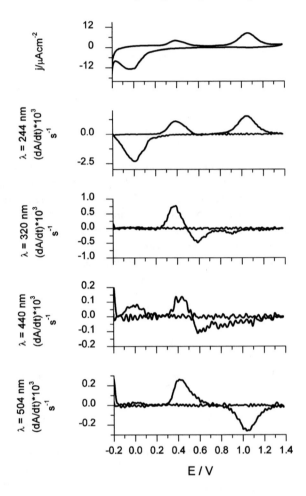

derivative of the absorbance and are presented as a function of the electrode potential $\left(\frac{dA}{dt}\right)$ vs E. These signals can be considered as an equivalent to voltammetric curves with absorption changes corresponding to the current flow (e.g. Bancroft et al. 1981; Zamponi et al. 1989).

Figure 3.5 shows typical cyclic voltammograms and voltabsorbommograms (VABs) recorded in 4 M H_2SO_4 containing 0.5 mM TcO_4^-. Chotkowski assumed that the wavelength of 244 nm represents TcO_4^-, 320 nm is attributed to some polymeric form of Tc(IV) (according to Vongsouthi 2009), 440 nm is probably due to Tc(III) while 502 nm is connected with polymeric Tc(IV) (e.g. Mausolf et al. 2011). The latter Tc species is probably the one referred to as electro-inactive Tc by Grassi et al. (1979). Moreover, Fig. 3.5 indicates that the above-mentioned Tc(III) species are the first pertechnetates electroreduction products, which are spectoscopically active under the experimentel conditions applied. This process manifests itself by formation of a positive spectroscopic wave (for 440 nm) at potentials below 0.2 V. The subsequent oxidation of such generated TcO_4^- electroreduction products leads to formation of some Tc(IV) species. Electrochemical oxidation of polymeric Tc(IV) forms starts at potentials higher than ca. 0.8 V as follows from an analysis of the wave at 502 nm. Chotkowski et al. (2018a) proposed the schemes of formation of these species as the products of mutual interactions between reduced Tc species, Eqs. (3.25)–(3.26):

$$TcO^+ + TcO^{2+} \rightarrow [Tc(\mu\text{–}O)_2Tc]^{3+} \qquad (3.25)$$

$$TcO^{2+} + TcO^{2+} \rightarrow [Tc(\mu\text{–}O)_2Tc]^{4+} \qquad (3.26)$$

or synproportionation of Tc(III) with Tc(VII) species, Eq. 3.27:

$$3TcO^+ + TcO_4^- + 6H^+ \rightarrow 2[Tc(\mu\text{–}O)_2Tc]^{4+} + 3H_2O \qquad (3.27)$$

Vichot (2001) discussed possible reversible reactions inside the polymeric technetium(III/IV) structures, Eq. (3.28):

$$Tc^{IV} - Tc^{IV} + e^- \leftrightarrows Tc^{III} - Tc^{IV} + e^- \leftrightarrows Tc^{III} - Tc^{III} \qquad (3.28)$$

This author observed that reduction of Tc(VII) with a concentration of μM carried out at -0.4 V versus Ag, AgCl(KCl(sat.)) in a solution with pH $= 1.75$ and containing $[SO_4^{2-}] = 0.01$ M leads to formation of a UV–Vis band at 497 nm, which is characteristic of polymeric Tc(IV) form. Longer than 1.3 h polarization at the same potential, up to 3.1 h, resulted in the disappearance of this band and formation of another one at 570 nm. This spectroscopic signal has been attributed to $Tc^{III} - Tc^{IV}$ forms with 3.5 as the formal charge of the Tc atom.

Other papers devoted to the technetium systems containing polymeric Tc(IV) species include works of Vongsouthi (2009). The reported experiments were carried out in $1 \div 9$ M HTFMS. This author stated that the Tc(IV) polymer is stabilized in strongly acidic solutions. The standard redox potential of this newly described polymeric Tc(IV)/Tc^{3+} couple was reported to be equal to 0.469 V versus NHE. When concentration of the pertechnetates is low (below ~ 0.5 mM), their electroreduction leads to the formation of Tc(III). When the technetium concentration increases above c.a. 0.2 mM a process of the Tc polymerization is observed. An increase in the acid concentration results also in stabilization of monomeric species containing Tc(III) and their reoxidation products, i.e., Tc(IV). The last step of the process comprises the formation of the Tc(III) as follows from the presence of an UV–Vis absorption band around $460 \div 480$ nm. Some of these Tc(III) may exist in polymeric forms.

Chotkowski and Czerwiński (2014a) studied also reduction of dimeric reduced technetium species. They investigated this process in strongly acidic media (4 M H$_2$SO$_4$) using UV–Vis spectroelectrochemistry with RVC-OTTLE electrodes. They

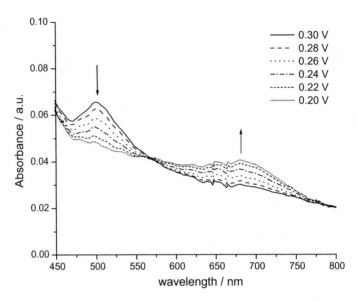

Fig. 3.6 UV–Vis spectra recorded during chronoamperometric reduction of $[Tc_2O_2]^{3+}$ in 4 M H_2SO_4 (reprinted with permission from Chotkowski and Czerwiński (2014b) Copyright 2014 Elsevier)

assumed the generation of $[Tc_2O_2]^{3+}$ at 0.3 V versus SHE. At potentials lower than 0.3 V, the intensity of the band at 500 nm decreased while above 600 nm an additional increase in the absorbance was observed (Fig. 3.6). These authors concluded that such evolution of the Vis spectrum is related to the generation of hydrated technetium(III) oxyhydroxide, which exists in the solution as $TcO(OH)_{aq}$. The overall reaction is given by Eq. (3.29):

$$[Tc_2O_2]^{3+} + 2H_2O + e^- \leftrightarrows 2TcO(OH)_{aq} + 2H^+ \qquad (3.29)$$

These authors proposed also the oxyhydroxide as the final reaction product on the basis of literature data but this conclusion should be confirmed using respective crystallographic techniques. An increase in the absorption band above ca. 600 nm is probably due to scattering of the light on the aggregates of hydrated technetium(III) oxyhydroxide (Fig. 3.6). Calculated standard redox potential of the analyzed Tc(IV)/Tc(III) couple is equal to 0.512 V versus SHE and is in a good agreement with the values reported by other authors for the Tc(IV)/Tc(III) system (Rard 1983).

So far, no solid evidence has been provided regarding the existence of Tc^{2+} in aqueous solutions. Early literature reviews (see e.g., Magee and Cardwell 1974) suggested the formation of these forms and they reported standard redox potential of, e.g., Tc^{2+}/Tc system as equal to $E^\ominus = 0.4$ V. However, more recent works do not recommend assumptions about the formation of Tc^{2+} species (Rard et al. 1999). Due to the fact that the metallic technetium is important in nuclear fuel reprocessing, its electrochemical properties are described in Chap. 5.

Chemistry of halides containing reduced technetium is one of the most extensively explored areas of the technetium chemistry. This is due to the fact that these compounds are important for various fields of the nuclear industry, including production of the nuclear fuel (e.g. volatile TcF_6) and management of radioactive waste (e.g. immobilization of poorly reactive compounds in solid forms). Technetium halides also found application as precursors of various radiopharmaceuticals.

Trop (1979) examined electrochemistry of halogen complexes of technetium. He analyzed [n–Bu$_4$N]$_2$[TcCl$_6$] irreversible reduction in CH_3CN with 0.1 M [n–Bu$_4$N][ClO$_4$] used as supporting electrolyte and determined respective half-wave potential value, $E_{1/2}$, as equal to -0.34 V versus SCE. Oxidation of this Tc compound was also irreversible with $E_{1/2} = 1.88$ V. The respective $E_{1/2}$ of bromide complexes of Tc, $TcBr_6^{2-}$, are lower than the above values and are equal to -0.27 V and 1.70 V, respectively. This behavior suggests that TcX_6^{2-} (X = Cl or Br) are kinetically labile.

Rajec i Macášek (1981) dealt with studies on electoreduction process of Tc(V) chlorocomplexes in 4 M HCl. The investigated Tc species, $TcOCl_5^{2-}$, were obtained as product of a reaction of TcO_4^- with 11.8 M HCl. The electroreduction of Tc(VII) at -0.2 V and -0.4 V versus Ag, AgCl leads to the formation of $TcCl_5(H_2O)^-$ and $TcCl_6^{2-}$. On the basis of UV–Vis spectroscopy measurements in 286–357 nm region, these authors determined the concentration ratio of selected Tc(V) forms, such as, e.g., $TcOCl_5^{2-}$ with $\lambda_{max} = 238, 325$ nm. The equilibrium constant (K) for the reaction given by Eq. (3.30):

$$TcCl_5^- + Cl^- \leftrightarrows TcCl_6^{2-} \tag{3.30}$$

was found to be equal to 0.80 ± 0.04 M^{-1} ($T \sim 25\ °C$).

The speciation of Tc(IV) in chloride solutions was of interest for Liu et al. (2005). Electrophoretic experiments allowed them determining mobility of selected Tc(IV) forms. In 1 M HCl/NaCl electrolyte with pH of 1 and at 25 °C, these values are equal to $5.47\cdot10^{-4}$ cm$^2\cdot$V$^{-1}\cdot$S^{-1} and $2.13\cdot10^{-4}$ cm$^2\cdot$V$^{-1}\cdot$S^{-1} for $TcCl_6^{2-}$ and [$TcCl_5(H_2O)$]$^-$ respectively.

The work of the Deutsch group broadened significantly knowledge about the electrochemical properties of technetium halide complexes (e.g. Huber et al. 1987, 1981). These authors examined chloro- and bromotechnetium compounds in concentrated HX/X$^-$ (X: Cl, Br) and nonaqueous solutions. Selection of type of the solution was driven by the fact that all TcX_6^{2-}, except TcF_6^{2-}, undergo a fast hydrolysis in neutral and weakly acidic aqueous solutions with formation of insoluble TcO_2.

Huber et al. (1987) also analyzed the process of the TcX_6^{2-} (X : Cl, Br) electroreduction in $2 \div 4$ M HX/NaX($2 \div 5.1$ M for Cl or $2 \div 4$ M for Br). The overall reaction of reduction of Tc(IV) to Tc(III) was given by Eq. (3.31):

$$Tc(IV) + e^- \leftrightarrows Tc(III) + n_hH^+ + n_XX^- \tag{3.31}$$

while the corresponding Nernst equation is given by (3.32):

Fig. 3.7 Thin layer cyclic
voltammograms recorded in
2 M HCl + 2 M NaCl
solution with (solid line) and
without (dashed line)
addition of of 3 mM
[NH$_4$]$_2$TcCl$_6$. Scan rate of
5 mV·s^{-1}, potential given in
respect of SSCE (saturated
sodium calomel electrode)
(reprinted with permission
from Huber et al. (1987)
Copyright 1987 American
Chemical Society)

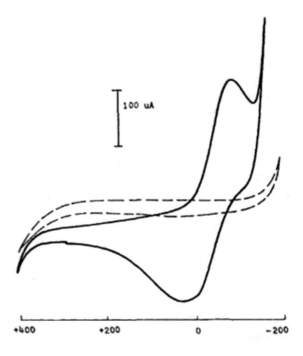

$$E = E^0 - 0.059(n_h) \log[H^+] - 0.059(n_X) \log[X^-] + 0.059\log\left(\frac{[\text{Tc(IV)}]}{[\text{Tc(III)}]}\right)$$

$$(3.32)$$

CV measurements with low rates of the working electrode potential scan (e.g. 5 mV·s^{-1}) reveal the existence of a redox couple (Fig. 3.7). The cathodic wave is broad and poorly shaped, which suggests that the cathodic reaction connected with this signal cannot be considered as a one-electron reversible process. An increase in the potential scan rate to 10 V·s^{-1} leads to a decrease in magnitude of the current peaks. It was concluded on the basis of these observations that the electrochemical reduction of Tc (IV) to Tc (III) is a slow process or, apart from the electrochemical step, this process may also include a slow chemical stage.

Huber et al. (1987) applied a nonlinear least-square fitting analysis to spectrophotometric data obtained for [Tc(IV)] = [Tc(III)]. The authors obtained the values of n_h, n_X in Eq. (3.31) and $E^{\ominus}$. For TcCl$_6^{2-}$ in HCl solutions: n_h= 1.2±0.1; n_X= 2.7±0.1; $E^{\ominus}$ = 0.082±0.004 V versus SSCE. For TcBr$_6^{2-}$ in HBr, these values were equal to n_h = 1.6±0.3; n_X= 5.9±0.5; $E^{\ominus}$= 0.240±0.017 V versus SSCE. These results indicate that the technetium bromide complexes are easier to reduce than those containing chloride. Further on, lower oxidation states of techentium are more stable in complexes with bromide as compared with the chloride containing ones. The TcCl$_6^{2-}$ ions electroreduction is accompanied by substitution of only three chloride ligands with water molecules while electroreduction of TcBr$_6^{2-}$ leads to the formation of bromide-free aquahydroxomplexes of Tc(III) according to the scheme

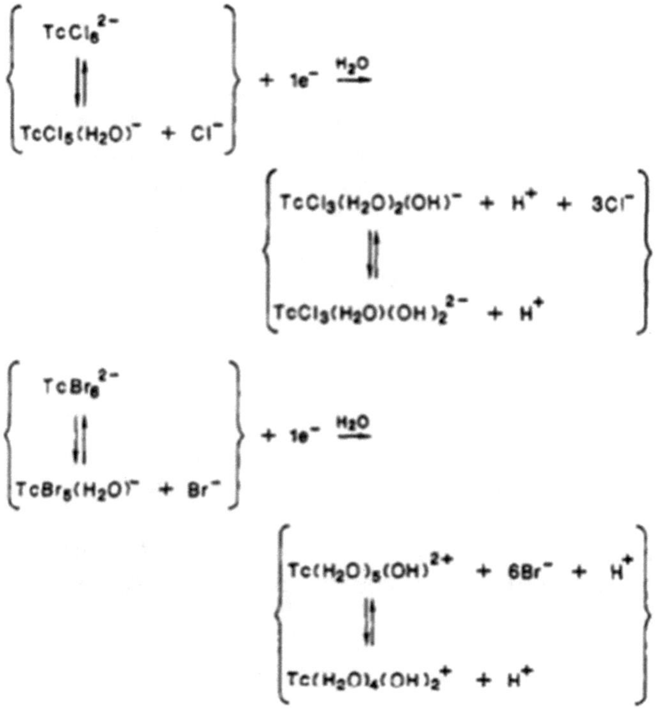

$$\left\{ \begin{array}{c} \mathrm{TcCl_6^{2-}} \\ \updownarrow \\ \mathrm{TcCl_5(H_2O)^-} + \mathrm{Cl^-} \end{array} \right\} + 1e^- \xrightarrow{\ H_2O\ }$$

$$\left\{ \begin{array}{c} \mathrm{TcCl_3(H_2O)_2(OH)^-} + \mathrm{H^+} + 3\mathrm{Cl^-} \\ \updownarrow \\ \mathrm{TcCl_3(H_2O)(OH)_2^{2-}} + \mathrm{H^+} \end{array} \right\}$$

$$\left\{ \begin{array}{c} \mathrm{TcBr_6^{2-}} \\ \updownarrow \\ \mathrm{TcBr_5(H_2O)^-} + \mathrm{Br^-} \end{array} \right\} + 1e^- \xrightarrow{\ H_2O\ }$$

$$\left\{ \begin{array}{c} \mathrm{Tc(H_2O)_5(OH)^{2+}} + 6\mathrm{Br^-} + \mathrm{H^+} \\ \updownarrow \\ \mathrm{Tc(H_2O)_4(OH)_2^{+}} + \mathrm{H^+} \end{array} \right\}$$

Fig. 3.8 Reduction schemes of $\mathrm{TcCl_6^{2-}}$ and $\mathrm{TcBr_6^{2-}}$ in 2 M HX (X = Cl, Br) (reprinted with permission from Huber et al. (1987) Copyright 1987 American Chemical Society)

shown in Fig. 3.8. Halide complexes of Tc(III) are more labile and unstable than those of Tc(IV). The authors stated that the half-life time for substitution of X by water in concentrated HX/X$^-$ media would appear to be in the range from 0.1 s. to few minutes.

Tetrabromo and tetrachloro oxo- and nitridotechnetium(V) complexes were later examined in nonaqueous solutions by Baldas et al. (1998). It turned out that, the axial O^{2-} or N^{3-} ligands have of great impact on observed redox potentials for Tc^{VI}/Tc^{V} and Tc^{V}/Tc^{IV} systems (Table 3.5).

Table 3.5 The half-wave potentials for selected oxo- and nitrido-technetium complexes (solution: 0.5 M (Bu$_4$N)PF$_6$ in dichloromethane, potential given versus SCE, Baldas et al. 1998)

Compound	$E_{1/2}$/V	
	Tc(V)/Tc(VI)	Tc(IV)/Tc(V)
[Bu$_4$N][TcOCl$_4$]	1.84	-0.52 $\left(E_p^c,\ \text{irrev}\right)$
[Bu$_4$N][TcOBr$_4$]	1.73 $\left(E_p^a,\ \text{irrev}\right)$	-0.39 $\left(E_p^c,\ \text{irrev}\right)$
[Bu$_4$N][TcNCl$_4$]	0.21	
[Bu$_4$N][TcNBr$_4$]	0.32	

One-electron redox couples were recorded for all examinated Tc(V)/Tc(VI) couples. Substitution of N^{3-} for O^{2-} in these types of technetium complexes results strong decrease in $E_{1/2}$. Based on the spectroelectrochemical results, the authors concluded that five-coordinate geometry of $[Tc(N \text{ or } O)X_4]^-$ is preserved (in contrast to aqueous solutions) during discussed redox reaction.

The problem of stability of oxochloro technetium (IV/III) complexes was of interest to Said et al. (2000). Oxochlototechnetates(2−) slowly decompose in 1 M HCl to $TcCl_mO_n(OH)_p(H_2O)_q^{(4-m-2n-p)+}$ where $m + n + p + q = 6$. Electrochemical measurements carried out in 5 M (H, Na) Cl solution showed that the reduction of the product of the decomposition of hexachlototechnetates (2−) may be given by Eq. (3.33):

$$TcCl_mO_n(OH)_p^{(4-m-2n-p)+} + (2n + p - 2r - s)H^+ + e^- \leftrightarrows$$

$$TcCl_qO_r(OH)_s^{(4-m-2n-p)+} + (m - q)Cl^- \tag{3.33}$$

where: $m > q, n > r, p > s$ and $(2n + p - 2r - s) > 0$. The equation describing redox potential for Tc(IV)/Tc(III) couple derived by the authors derived has the following form (Eq. 3.34):

$$E_{mp} = E'^0 + (2n + p - 2r - s) \cdot 0.059 \log[H^+]. \tag{3.34}$$

where E_{mp} represents the potential value for $[Tc(IV)] = [Tc(III)]$. Based on the data given by Said et al. (2000), the standard redox potential of Tc(IV)/Tc(III) couple (Eq. (3.33)) can be estimated as located within the $0.016 \div 0.029$ V versus Ag/AgCl range for 25 °C.

Hurst et al. (1981) investigated the stability of isothiocyanate Tc(IV) complexes in an acetonitrile solution containing 0.1 M tetrabutylammonium perchlorate. The half-wave potential of $[Tc(NCS)_6]^{2-}/[Tc(NCS)_6]^{3-}$ couple determined by them on the basis of polarographic results is equal to 0.18 V versus SCE.

Information on the physico-chemical properties of technetium cluster compounds can be found in several reviews (Kryuchkov et al. 1979; German and Kryutchkov 2002; Sattelberger 2005). Synthesis and spectroscopic characterization of new technetium halides is one of the fields of research of Poineau (see e.g., Poineau et al. 2014).

Kryutchkov (1999) presented a general scheme of the pertechnetates reduction in solutions containing hydrohalogenic acid. The proposed reactions are summarized in Eq. (3.35):

$$TcO_4^- \rightarrow TcOX_4^- \rightarrow TcX_6^{2-} \leftrightarrows [Tc_2X_8]^{2-} \leftrightarrows \cdots \leftrightarrows [Tc_xX_y]^{n-} \tag{3.35}$$

Technetium ions bond together to form large clusters with an increasing size. x in these clusters can be as large as 6 or 8 while y can reach even 14.

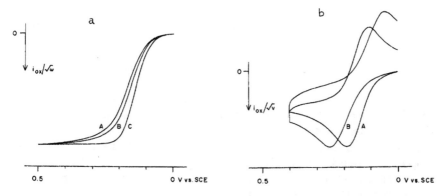

Fig. 3.9 a Rotating-disk electrode polarograms of 10^{-4} M Y[Tc$_2$Cl$_8$] 9H$_2$O in 10% v/v 12 M aqueous hydrochloric acid in ethanol using a platinum electrode and potential sweep rate of 2 mV·s^{-1}:(A) 30 rotations·s^{-1}; (B) 10 rotations·s^{-1}; (C) 1 rotations·s^{-1}. **(b)** Cyclic voltammetry in the same solution: (A) 5 mV·s^{-1}; (B) 200 mV·s^{-1} (reprinted with permission from Cotton and Pedersen (1975) Copyright 1975 American Chemical Society)

Cotton and Pedersen (1975) analyzed the electrochemical properties of octachloroditechnetate(2− or 3−). [Tc$_2$Cl$_8$]$^{3-}$ was obtained in a reaction of TcCl$_6^{2-}$ with zinc in concentrated HCl solutions at 70 °C. In a solution containing 10% v/v 12 M HCl and ethanol, the reaction given by Eq. 3.36 is quasireversible with $E_{1/2}$= 0.140 versus SCE at 25 °C (Fig. 3.9):

$$[Tc_2Cl_8]^{2-} + e^- \leftrightarrows [Tc_2Cl_8]^{3-} \tag{3.36}$$

An analysis of polarograms recorded under hydrodynamic conditions for the working electrode rotation rates from 0.5 to 30 s^{-1} shown that $i_L \cdot \omega^{-1/2}$ was independent of (Fig. 3.9a). The limiting value of the current for the slope of 60 ± 1 mV, which is expected for a reversible reaction, was reached for rotation rates less than 3 s^{-1} and for v of 2 mV·s^{-1}. Moreover, the plots of $E = f\left(\log \frac{i_L - i}{i}\right)$ reveal the slopes varying from 80 to 60 mV when the rotation rates decrease from 30 to 0.5 s^{-1}, respectively. An analysis of cyclic voltammograms (Fig. 3.9b) reveals that the separation of the anodic and the cathodic peak potentials varies from 210 to 70 mV for 200 and 5 mV·s^{-1}, respectively. A constant value of i_p^c / i_p^a leads the authors to the conclusion that the lifetime of the oxidized form of dinuclear technetium, [Tc$_2$Cl$_8$]$^{2-}$, is longer than 300 s. The authors applied Nicholson's theory of the quasi-reversible reactions (Nicholson 1965) in analysis of the discussed redox technetium couple and they obtained value of $\Psi v^{1/2}$ equals to 0.08 ± 0.01 V$^{1/2}$·s$^{-1/2}$.

It should be stressed that dinuclear technetium clusters can react with oxygen and water and their stability depends also on acids concentration. Kryutchkov (1999) discussed possibility of the following reactions of [Tc$_2$Cl$_8$]$^{3-}$ in hydrohalogenic acid, e.g. Equations (3.37)–(3.39).

$$[Tc_2Cl_8]^{3-} + H_2O \leftrightarrows [Tc_2Cl_8(H_2O)_n]^{(3-n)-} \tag{3.37}$$

$$[Tc_2Cl_8]^{3-} + Cl^- \leftrightarrows [Tc_2Cl_9]^{4-} \qquad (3.38)$$

$$[Tc_2Cl_9]^{4-} + HCl \leftrightarrows Tc(II) + Tc(III) \qquad (3.39)$$

$[Tc_2Cl_8]^{3-}$ is relatively stable in $3 \div 6$ M aqueous HCl while $[Tc_2I_8]^{3-}$ decomposes in aqueous HI almost *in statu nascendi*. The acid concentration window for which these ions are stable becomes narrower according to the following order: HCl > HBr > HI. As compared with the aqueous solutions, dinuclear techentium compounds are much more stable in organic solvents.

Octachloroditechnetate(2- or 3-) ions undergo decomposition in the presence of oxygen according to reactions (3.40)–(3.41):

$$[Tc_2Cl_8]^{3-} \overset{O_2}{\rightarrow} [Tc_2Cl_8O_2]^{3-} \rightarrow [TcO(OH)Cl_4]^{2-} + TcCl_6^{2-} \qquad (3.40)$$

$$[Tc_2Cl_8]^{2-} \overset{O_2}{\rightarrow} [Tc_2Cl_8O_2]^{2-} \rightarrow [TcOCl_4]^- \qquad (3.41)$$

Armstrong and Taube (1976) analyzed the electrochemical properties of *trans*-aquonitrosyltetraamminetechnetium(I) ions, $[Tc(NH_3)_4(NO)H_2O]^{3+}$. In strong acidic solution of HTFM and NaAcO, a one-electron reversible couple was observed with $E_{1/2} = 0.8$ V versus NHE, Eq. (3.42):

$$trans-[Tc(NH_3)_4(NO)OH]^{2+} + H^+ + e^- \leftrightarrows trans-[Tc(NH_3)_4(NO)H_2O]^{2+}$$
$$(3.42)$$

This reversible system was observed in solutions containing high concentration of acid with pH not higher than 2. The replacement of water molecule by OH^- ions is rapid and does not involve Tc–O bond breaking. Additionally, the replacement of the NH_3 molecule by π-reach organic acid ligands, such as 1,10-phenantroline, results in the formation of a technetium complex, which exhibits an increased stability against oxidation. This process was observed for $[Tc(phen)_2(NO)NH_3]^{2+}$ but suprisingly not for monophenantroline complex for which $E_{1/2} = 0.69$ V. Moreover, the authors stated that nitrosylamminetechnetium(I) complex is easier to oxidase than its isoelectronic analog of Ru(II) whose oxidation to Ru(III) is observed at $E < 1.1$ V. Based on the electrochemical measurements, Armstrong and Taube estimated the pKa for dissociation of *trans*-$[Tc(NH_3)_4(NO)H_2O]^{3+}$ at a level of 2.

Also other nitrosyl complexes of technetium were investigated electrochemically (Balasekaran et al. 2014). The reduction of the pertechnetates with acetohydroxamic acid in aqueous HF leads to the formation of pentafluoronitrosyltechnetium(I and II). A well-developed one-electron redox system with $E_{1/2} = 0.652$ V (vversus NHE) was observed for $[Tc(NH_3)_4(NO)F]^+$ in 0.1 M KF/H$_2$O. The value of the separation of anodic and cathodic peaks, ΔE_p, was equal to 107 mV and did not reveal electrochemical reversibility of the system. An earlier study of Armstrong and Taube (1976) on *trans*-$[Tc(NH_3)_4(NO)F]^{2+}$ in aqueous solutions of trifluorementhanesulfonate acid showed $E_{1/2} = 0.8$ V versus NHE. This potential value was constant up to pH of

2. The authors stated that this redox couple is electrochemically reversible in strongly acidic solutions. At pH between 2 and 5, the slope in the plot of pH versus $E_{1/2}$ was equal to 35 mV per pH unit.

3.2 Alkaline solutions

Early works on reduction of the pertechnetates in alkaline solutions reported a multi-step process although the identity of the initial stage of the process is disputable. For example, Salaria et al. (1963b) observed a wave at a half-wave potential of -0.81 V versus SCE in a solution with pH of 13, which was attributed to an irreversible three-electron process. Further reduction of such generated technetium species to Tc(III) occurred at $E_{1/2} = -1.02$ V and was identified as an adsorption-controlled process. Münze (1968), on the other hand, reported a four-electron process composed of three stages. It begins with a transfer of two electrons (Tc(VII) → Tc(V)) at $E_{1/2} = -0.8$ V which is followed by a one electron reaction (Tc(V) → Tc(IV)) at $E_{1/2} = -0.9$ V and ends with another one electron step (Tc(IV) → Tc(III)) at $E_{1/2} = -1.1$ V versus SCE. In contrast, Colton et al. (1960) concluded that the pertechnetates are reduced in 0.1 M KOH to Tc(IV) and not to Tc(III). This process includes two- ($E_{1/2} = -0.85$ V versus SCE) and one-electron reduction steps ($E_{1/2} = -1.15$ V).

The electroreduction of the pertechnetates in alkaline solutions is assumed nowadays as a multistage process with Tc(V) as the product generated through Tc(VI) intermediates. The Tc(V) species can be further reduced to Tc(IV). It cannot be ruled out that electrolysis of a solution containing Tc(IV) at strongly cathodic potentials can lead to the formation of Tc(III) species and, under specific conditions, even metallic Tc can be generated.

The hexavalent technetium species and its stability were examined using UV–Vis by Deutsch et al. (1978). These researchers conducted pulse radiolysis of 0.01 ÷ 0.1 mM TcO_4^- in 0.1 M KOH solutions. They observed that both aqueous electron and TcO_4^- disappear according to a first-order kinetics. This process was accompanied by generation of a new form of Tc, which was described the researchers as possibly TcO_4^{2-}. They found the rate constant of this process equals to $k_2 = (2.48 \pm 0.05) \cdot 10^{10}$ $dm^3 \cdot mol^{-1} \cdot s^{-1}$ at 25 °C. The Tc(VI) form was characterized by two bands: the first one with a maximum at approximately 335 nm and the second one, with a broad, flat peak in the range of 500 ÷ 530 nm. Deutsch et al. stated that Tc(VI) is apparently stable for 10 ms but later on its reactions accelerate and all the species eventually disappears within 50 ms. In addition, cyclic voltammetry experiments (Fig. 3.10) revealed that a reduction wave at a potential of about -0.8 V versus SCE appears only for very high scan rates of the working electrode (Fig. 3.10b, c). The researchers attributed this wave to the Tc(VII)/Tc(VI) couple. Application of Randles–Sevcik equation in analysis of the voltammetric results allows to calculate number of the electrons exchanged in the discussed process, which was equal to $n = 1$. The standard reduction potential of the Tc(VII)/Tc(VI) system was determined at a level of -0.61 V versus NHE.

Fig. 3.10 Cyclic voltammograms resulting from reduction of TcO_4^-. Conditions: 25 °C; $[OH^-]$ = 0.12 M; $[TcO_4^-] = 1 \div 7 \cdot 10^{-3}$ M; nitrogen saturated solution; hanging mercury drop electrode, surface area $= 9.4 \times 10^{-3}$ cm^2. Curves (A), (B), and (C) result from scan rates of 1, 20, and 100 V·s^{-1}, respectively (reprinted with permission from Deutsch et al. (1978) Copyright 1978 American Chemical Society)

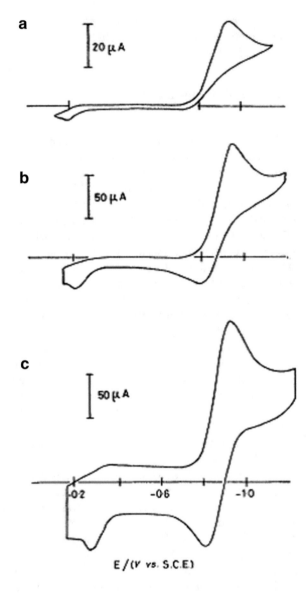

Astheimer et al. and Schwochau et al. (Astheimer and Schwochau 1976; Schwochau et al. 1974) pointed out that TcO_4^{2-} ions are extremely sensitive to the presence of oxygen and atmospheric moisture. Under such conditions, the Tc(VI)) undergoes oxidation and disproportionation reactions. These researchers investigated the process of TcO_4^- ion reduction in acetonitrile with $(CH_3)_4NClO_4$ as a supporting electrolyte. The half-wave potential for a one-electrode Tc(VII)/Tc(VI) redox couple was found to be equal to $E_{1/2} = -1.74$ V versus SCE.

It is worth mentioning here that the knowledge about the structure of the Tc(VI) species that are present in an aqueous environment is incomplete and ideas of various researchers are often inconsistent.

Kissel and Feldberg (1969) investigated the pertechnenates reduction in 1 M NaOH. They found that disproportionation) of Tc(VI) is very fast and competes with generation of Tc(V). This process is presented by Eq. (3.43):

$$2TcO_4^{2-} \rightarrow TcO_4^- + TcO_4^{3-} \tag{3.43}$$

The respective rate constant determined by these authors is equal to 1.5×10^5 $dm^3 \cdot mol^{-1} \cdot s^{-1}$. The Tc(VI) disproportionate with a much faster rate than the Tc(V) whose rate constant is equal to 2.4×10^3 $dm^3 \cdot mol^{-1} \cdot s^{-1}$ (see: Lukens et al. 2001). Kissel and Feldberg (1969) concluded that the overall reduction of the pertechnetates in alkaline media is a two electron process.

Nowadays, the recommended value of the standard redox potential for the Tc(VII)/Tc(V) couple, Eq. (3.44):

$$Tc(VII) + 2e^- \leftrightarrows Tc(V) \tag{3.44}$$

is equal to -0.60 ± 0.05 V (Rard et al. 1999) being only slightly higher than the redox potential for Tc(VII)/Tc(VI) couple, Eq. (3.45), which is equal to $E^{\ominus} = -0.64 \pm 0.03$ V:

$$Tc(VII) + e^- \leftrightarrows Tc(VI) \tag{3.45}$$

Kryuchkov et al. (1979) determined the protonation constants of the following reactions of the technetates(VI) (Eqs. (3.46) and (3.47)):

$$TcO_4^{2-} + H^+ \leftrightarrows HTcO_4^- \tag{3.46}$$

$$HTcO_4^- + H^+ \leftrightarrows H_2TcO_4 \tag{3.47}$$

$\log_{10}K_1$ of reaction (3.46) is equal to 8.7 ± 0.5 while for the process (3.47) $\log_{10}K_1 \leq 1$. Accuracy of determination of these values is hard to evaluate due to instability of the discussed Tc species.

The values of the protonation constant of the pertechnetates (reaction (3.48)) given by various authors are inconsistent and reported $\log_{10}K$ varies from -0.39 to 0.60 (see: Rard et al. 1999, pp. 101–102), Eq. (3.48):

$$TcO_4^- + H^+ \leftrightarrows HTcO_4 \tag{3.48}$$

Regardless on these discrepancies, one may assume that these values are high enough as to assume that the $HTcO_4$ is fully dissociated in aqueous solutions.

Founta et al. (1987) proposed a scheme of the processes of the pertechnetates reduction in alkaline media (0.01 ÷ 0.1 M NaOH), which includes a one electron electrochemical reduction of the Tc(VII) (reaction (3.49)). In a subsequent step, the product of this reaction, TcO_4^{2-} ions undergo a hydrolysis, according to Eqs. (3.50) and (3.51):

$$TcO_4^- + e^- \leftrightarrows TcO_4^{2-} \tag{3.49}$$

$$TcO_4^{2-} + H_2O \leftrightarrows TcO_3(OH)^- + OH^- \tag{3.50}$$

$$TcO_3(OH)^- + H_2O \leftrightarrows (TcO)O(OH)_3^- \tag{3.51}$$

An interaction of nonhydrolyzed technetates(VI) ions with hydrolyzed Tc(VI) leads to the formation of hydrolyzed Tc(V) species, according to Eqs. (3.52) and (3.53):

$$TcO_4^{2-} + (TcO)O(OH)_3^- \leftrightarrows TcO_4^- + (TcO)O(OH)_3^{2-} \tag{3.52}$$

$$(TcO)O(OH)_3^- + e^- \leftrightarrows (TcO)O(OH)_3^{2-} \tag{3.53}$$

These authors concluded that in contrast to synproportionation of Tc(V) and Tc(VII), the disproportionation) process of Tc(VI) must be very fast. The rate-determining step of this mechanism is an addition of one water molecule to $TcO_3(OH)^-$, which leads to generation of $(TcO)O(OH)_3^-$ containing 5-coordinated Tc, Eq. (3.51). It should be stressed that coordination of two water molecules to $TcO_3(OH)^-$, which leads to the formation of six-coordinated form of $(TcO_2)O(OH)_3OH_2^-$ is also possible. Incoming ligand, H_2O, binds to the technetium core and transfers H^+ to an oxyl group. Apart from the disproportionation of Tc(VI) described by the Eq. (3.52), the electrochemical reduction of $(TcO)O(OH)_3^-$ to $(TcO)O(OH)_3^{2-}$ also can occur (Eq. 3.53). The authors also stated that at pH above 12 the Tc(V) can be reduced to Tc(IV) with an unknown structure. The kinetics of this reaction is of second order in respect to Tc.

Founta et al. (1987) reported an influence of the Tc concentration on the calculated number of the electrons exchanged in the Tc(VII) reduction in 0.1 M NaOH. This value was constant and equal to 2.10 for the TcO_4^- concentration below ca. 0.1 mM but for higher concentrations it increased significantly reaching ca. 2.7 for [Tc] of ca. 1 mM. Such behavior suggests possible dimerization of the reduced technetium species at higher Tc concentrations. These results are in line with other literature reports dealing with studies on polymerization of Tc(IV) in other solutions (see: e.g. Vichot et al. 2003; Poineau et al. 2006).

Chatterjee et al. (2018) analyzed electroreduction of pertechnetates in NO_3^-/NaOH solutions using the scheme proposed by Founta (3.49–3.52). They assumed Tc(VI) (or, alternatively TcV)) as the final product of this reaction in high

Table 3.6 Simulation parameters presented in Chatterjee et al. (2018) describing hydrolysis and disproportionation) of Tc(VI) species electrogenerated in 5 mM $TcO_4^- + NaNO_3/2$ M NaOH

Reaction	Parameter		
	K_{eq}	k_f	k_b
$TcO_4^{2-} + H_2O \leftrightarrows TcO_3(OH)^- + OH^-$	10^7	1.8	1.8×10^{-7}
$TcO_3(OH)^- + H_2O \leftrightarrows (TcO)O(OH)_3^-$	10^7	0.033762	3.3762×10^{-9}
$TcO_4^{2-} + (TcO)O(OH)_3^- \leftrightarrows TcO_4^- + (TcO)O(OH)_3^{2-}$	10^8	2×10^4	2×10^{-4}

ionic strength media. Spectroelectrochemical experiments showed that the reduction of TcO_4^- leads to formation of a wave at 440 nm, which was attributed by these authors to generation of Tc(VI). Simultaneous appearance of a shoulder at 657 nm may indicate the presence of also Tc(V). Poineau et al. (2013) reported that in strongly acidic media the Tc(V) can be characterized by a weak band above 600 nm. An analysis of the standard potential of a redox couple that involves Tc(VII) ($E^\ominus = -0.819$ V mV versus Ag, AgCl) in combination with the respective number of exchanged electrons ($n_e = 0.9$) determined by Chatterjee and coworkers indicates that under applied experimental conditions, the Tc(VII) is reduced to Tc(VI), Eq. (3.49). Calculated values of the respective rate constant (k_s) and transfer coefficient (α) are equal to 8.1351×10^{-3} s·cm^{-1} and 0.35, respectively. The k_s is about 10 times smaller than for example a reversible system of $Fe(CN)_6^{3-}/Fe(CN)_6^{4-}$ ($\Delta E_p = 59$ mV, $k_s = 9 \times 10^{-2}$ s·cm^{-1}, $\alpha \sim 0.5$) but is comparable to a quasi-reversible Fe^{3+}/Fe^{2+} couple ($\Delta E_p > 59$ mV, $k_s = 5.3 \times 10^{-3}$ s·cm^{-1}, $\alpha \sim 0.4$) (see: Galus 1976; Li et al. 2016).

Apart from the spectroscopic data, Chatterjee et al. (2018) also analyzed cyclic voltammograms using numerical simulations and assuming the reactions scheme proposed by Founta (1987) (reactions (3.50)–(3.52)). They determined the kinetic parameters of these reactions (Table 3.6). The reported simulations do not reproduce completely correctly experimental results in the whole potential range. However, it must be stressed that such calculated values of the kinetic parameters of the hydrolysis and disproportionation of Tc(VI) in alkaline solutions are the one of the few reported in the literature.

The rate-determining step of the Tc(VI) disproportionation) is the hydrolysis of $TcO_3(OH)^-$, which produces $(TcO)O(OH)_3^-$. Unstable Tc(VI) forms undergo a transformation, which follows a first-order kinetics with unexpectedly long half-life equal to 1.98 ± 0.09 day.

The electroreduction of the pertechnetates in strongly alkaline solutions (0.3 ÷ 10.6 M NaOH) was analyzed by Chotkowski et al. (2018a). These authors observed the formation of two bands at 460 and 495 nm, which were attributed to generation of Tc(V) and Tc(IV) species, respectively. Dimeric forms of Tc(IV) ($\lambda = 495$ nm) are generated in an extremely strong alkaline environment. In contrast to the reports of Chatterjee et al. (2018), these researchers did not observe the formation of the band at 440 nm. It should be stressed that the origin of the band at 400÷500 nm is still unclear. Identification of the wave near 465 nm as the one attributed to Tc(V) can be supported by results obtained in acidic solutions by Rotmanov et al. (2015) but is in contrast to conclusion of Poineau et al. (2013) who connected the Tc(V) to a significantly different wavelength. Noteworthly is the fact that formation of insoluble Tc(IV) species leads to an overall increase in the UV–Vis absorbance which may mask the presence of Tc forms whose absorbance is very low. The number of the electrons involved in the electrochemical reduction of the pertechnetates determined by Chotkowski et al. varies from 1.5 to 1.7 being higher than that reported by Chatterjee et al. According to the former authors, this effect confirms the occurrence of the Tc(VI) disproportion. Additional Au-SERS measurements revealed the formation of intermediate Tc forms, probably Tc(V), which exhibits a band at 900 cm^{-1}. The shape of the CV curves recorded for TcO$_4^-$ in alkaline solutions by Chotkowski et al. is very similar to those reported by Chatterjee et al. Both groups observed a single reduction wave during the cathodic potential scan and two main oxidation current waves recorded after reversing direction of the potential changes. In NaOH solutions the latter signals were observed at ca. $-0.7 \div -0.6$ V and $-0.4 \div 0$ V (versus Hg, HgO (0.1 M NaOH)) (Fig. 3.11). The shape of these peaks strongly depends on the NaOH concentration. In less alkaline solutions (0.3 M NaOH), the first of these signals is definitely less visible than in concentrated ones, such as e.g. 10 M. Such evolution of the current peak indicates the stabilization of the intermediate Tc(V) and Tc(VI) species in concentrated alkaline solutions. This is in line with conclusions of

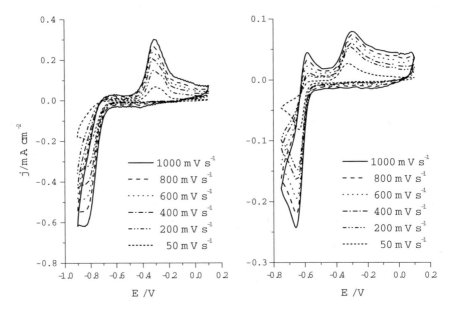

Fig. 3.11 Cyclic voltammograms recorded with various potential sweep rates for an Au electrode in 1 mM KTcO$_4$ + 0.3 M NaOH (left panel) and 10 M NaOH (right panel), $E_{start} = 0.1$ V (reprinted with permission from Chotkowski et al. (2018b) Copyright 2018 Elsevier)

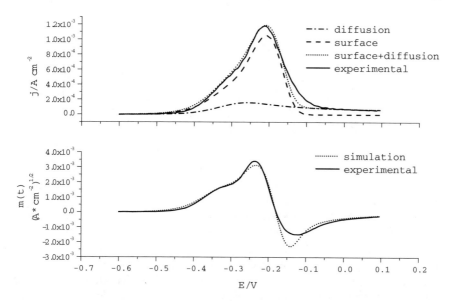

Fig. 3.12 Linear sweep voltammograms (top) of an Au electrode in 0.5 mM KTcO$_4$ + 0.3 M NaOH recorded with 1 V s^{-1} after chronoamperometric reduction of the pertechnetates at −0.65 V, and their semiderivatives (bottom). E_{start} = −0.6 V, (reprinted with permission from Chotkowski et al. (2018b) Copyright 2018 Elsevier)

Chatterjee et al. who reported an increase in stability of the Tc(VI) in solutions with a high ionic strength.

Chotkowski et al. (2018b) also examined the oxidation of reduced Tc species deposited/adsorbed on a gold electrode in 0.3 M NaOH. On the basis of analysis of CVs and semiderivatives of currents due to oxidation of reduced Tc (Fig. 3.12), they concluded a simple electrochemical model which did not include disproportion of Tc(V) and Tc(VI) but follows two parallel oxidation pathways:

1. "surface" pathway, starting with the electrode surface covered with Tc(IV) or Tc(V), scheme 3.54:

$$\text{Tc(IV)}_{(s)} \xrightarrow{k^0(1),\, E^0(1), \alpha(1)} \text{Tc(V)}_{(s)} \xrightarrow{k^0(2),\, E^0(2),\, \alpha(2)} \text{Tc(VI)}_{(s)}$$
$$\xrightarrow{k^0(3),\, E^0(3),\, \alpha(3)} \text{Tc(VII)} \tag{3.54}$$

2. "diffusion" pathway, that starts with soluble Tc(IV) present in the electrolyte, scheme 3.55:

$$\text{Tc(IV)}_{(d)} \xrightarrow{k^0(4),\, E^0(1),\, \alpha(4),\, D(4)} \text{Tc(V)}_{(d)} \xrightarrow{k^0(5),\, E^0(2),\, \alpha(5),\, D(5)} \text{Tc(VI)}_{(d)}$$
$$\xrightarrow{k^0(6),\, E^0(3),\, \alpha(6),\, D(6)} \text{Tc(VII)} \tag{3.55}$$

Table 3.7 The values of the parameters corresponding to the processes describing oxidation of Tc(IV) in 0.3 M NaOH (Chotkowski et al. 2018b)

Parameter	Step					
	1	2	3	4	5	6
$k^0/s^{-1} \cdot cm^{-2}$	0.22	0.004	>0.5 (even as high as 2)	0.00072	1.8×10^{-4}	$1.6 \div 0.02$
α		0.45	0.4	0.5	0.5	0.5	0.5
E^0/V	-0.52	-0.695	-0.805			
$D/cm^2 \cdot s^{-1}$				1.0×10^{-5}	1.2×10^{-5}	1.2×10^{-5}

The authors calculated basic electrochemical parameters of the reactions in question, such as the rate constants (k^0) and transfer coefficients (α). It turned out that the fastest step is the one related to the oxidation of Tc(VI) to Tc(VII) with k^0 equal to at least $0.5 \ s^{-1} \cdot cm^{-2}$ and possibly as high as $2 \ s^{-1} \cdot cm^{-2}$. Oxidation of ionic forms of Tc(IV) and Tc(V) at the electrode surface, on the other hand, proved to be the slowest steps with k^0 equal to $0.00072 \ s^{-1} \cdot cm^{-2}$ and $1.8 \cdot 10^{-4} \ s^{-1} \cdot cm^{-2}$, respectively. The authors concluded that due to a huge number of the parameters describing the reactions (3.54) and (3.55), which are used as variable in the fitting procedure it is possible to find many sets of the variables with different values which lead to a similar quality of the fit. The values of the fitted parameters proposed by the authors are given in Table 3.7.

Initial coverage of the electrode surface by Tc(IV)$_{(s)}$ and Tc(V) $_{(s)}$: $c(1) = 4.5 \times 10^{-10}$ mol·cm^{-2} and $cs(2) = 1 \times 10^{-10}$ mol·cm^{-2} respectively; initial concentration of Tc(IV)$_{(d)} = 1.95 \times 10^{-5}$ mol·dm^{-3}.

One of the most important issues related to the technetium chemistry is speciation of Tc(IV) in aqueous solutions in the presence of bicarbonates. This process can be strongly complicated by hydrolysis and complexation of the technetium containing species. A scheme of electrochemical reactions involved in these processes was proposed by Paquette and Lawrence (1985) and by Alliot et al. (2009).

Paquette and Lawrence (1985) investigated redox chemistry of Tc(IV)/Tc(III) couple in trifluoromethanesulfonate (NaTFMS) and sodium bicarbonate (pH = 8, I = 1) solutions by means of spectroelectrochemical methods. NaTFMS was selected as a supporting electrolyte due to instability of nitrates and perchlorates in the presence of reduced technetium species. The solutions were purged with a gas containing a mixture of carbon dioxide and argon. Láng and Horányi (2003) reported that perchlorates are decomposed during deposition of technetium layers (this aspect of the technetium chemistry is discussed in more detail in Chap. 5). Paquette and Lawrence (1985) noted that the electroreduction of the pertechnetates at potentials lower than -0.7 V versus SCE leads to a change in the color of the solution from initially colorless to pink (Tc(IV)) to eventually pale blue. The latter color is attributed to the carbonate complexes of Tc(III) (λ_{max} = 470 and 630 nm) (Fig. 3.13). This form of technetium turned out to be very sensitive to the presence of oxygen.

Paquette and Lawrence proposed the following general equation, which presents the transformation of Tc(IV) into Tc(III), Eq. (3.56):

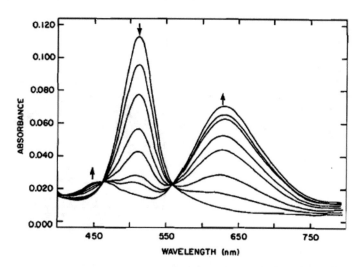

Fig. 3.13 Visible spectra recorded at successively decreased electrode potential from -0.5 V to -0.78 V (vs. SCE, note added by the authors) in a solution containing 1.2×10^{-4} mol dm^{-1} Tc(IV) solution with pH of 8, in 0.5 M "CO," medium. Arrows indicate peaks evolution with decreasing potential (Paquette and Lawren 1985, reprinted with permission from Paquette and Lawren (1985) Copyright 1985 Canadian Chemical Society)

$$\text{Tc(OH)}_p(\text{CO}_3)_q^{4-p-2q} + \text{H}_2\text{O} + e^- \leftrightharpoons \text{Tc(OH)}_{p+1}(\text{CO}_3)_q^{3-(p+1)-2q} + \text{H}^+ \quad (3.56)$$

where $p + 2q > 4$. The Tc(IV)/Tc(III) couple is considered as reversible or quasi-reversible one with $E^\ominus = -0.625$ V versus SCE.

More recent results of studies on the speciation of Tc(IV) in bicarbonate media ($[CO_3^{2-}] = 0.3 \div 1$ M; pH $= 7 \div 10$) were reported by Alliot et al. (2009). The potential of the redox system containing Tc(VII) and Tc(IV) in the bicarbonates solutions is described by the Eq. (3.57):

$$E = E^{0\prime}_{\text{TcO}_4^-/\text{TcO(OH)}_2} + \frac{0.059}{3}\left(\log\frac{[\text{TcO}_4^-][\text{H}^+]^4}{[\text{Tc(IV)}]} + \log\alpha_{\text{TcO(OH)}_2}\right) \quad (3.57)$$

where the Ringbom coefficient (for $\alpha_{\text{TcO(OH)}_2}$) can be written as, Eq. (3.58):

$$\alpha_{\text{TcO(OH)}_2} = \frac{[\text{Tc(IV)}]}{[\text{TcO(OH)}_2]} = 1 + K_1[\text{H}^+]^2 + K_2[\text{H}^+]$$
$$+ \frac{K_3}{[\text{H}^+]} + \sum_{q,p} K_{p,q}[\text{H}^+]^{4-p}[\text{CO}_3^{2-}]^q \quad (3.58)$$

The K_1, K_2 and K_3 constants can be taken from Table 3.3 The $K_{p,q}$ describes reaction (3.59) and can be calculated using Eq. (3.60):

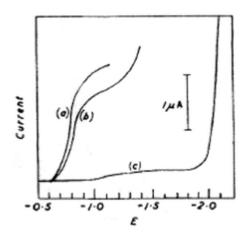

$$(4 - p)H^+ + TcO(OH)_2 + qCO_3^{2-} \leftrightarrows Tc(CO_3)_q(OH)_p^{4-2q-p} + (3 - p)H_2O \tag{3.59}$$

$$K_{p,q}^0 = K_{p,q} \frac{a_{H_2O}^{3-p} \gamma_{Tc}(CO_3)_q(OH)_p^-}{\gamma_{H^+}^{4-p} \gamma_{CO_3^{2-}}^{2-q}} \tag{3.60}$$

The standard redox potentials of $TcO_4^-/Tc(CO_3)(OH)_2$ and $TcO_4^-/Tc(CO_3)(OH)_3^-$ couples have been determined by means of potentiometric measurements and are equal to 733 ± 44 mV and 575 ± 60 mV versus SHE, respectively.

The electroreduction of pertechnetates in 0.1 M KCN was investigated by Colton et al. (1960). In this supporting electrolyte, the TcO_4^- ions gave a single well-developed irreversible polarographic wave with a half-wave potential of -0.81 V versus SCE. The calculated number of exchanged electrons was equal to 3 (Fig. 3.14).

Other authors, e.g., Trop et al. (1980a), synthesized cyanido complexes of technetium, e.g., heptacyanotechnetate(III) and oxopentacyanotechnetate(V). Such cyanide complexes have been less extensively investigated as compared with, e.g., halide complexes of Tc. $Tc(CN)_7^{4-}$ was prepared by a reaction of TcI_6^{2-} with KCN in methanol. This complex decomposes slowly in oxygenated water to other technetium(V) complexes, $TcO(CN)_5^{2-}$ and $trans-TcO_2(CN)_4^{3-}$. Results of studies on rhenium analogs shed some light on mobility of these Tc complexes in liquids. It is then expected that the diffusion coefficient values of these Tc complexes should be close to that reported for respective Re analogs, i.e., 0.55×10^{-5} cm$^{-2}\cdot$s^{-1} for $K_3Re(CN)_8$ in 0.1 M KCl and 0.7×10^{-5} cm$^{-2}\cdot$s^{-1} for $[ReO_2(CN)_4]^{3-}$ (see: Colton et al. 1960).

Electrochemical properties of octahedral hexatechnetium(III) clusters, $[Tc_6Q_8(CN)_6]^{4-}$ (Q = S, Se), were described by Yoshimura et al. (2010). These compounds were synthesized as the products of an axial halides substitution with cyanide in $Cs_4[Tc_6S_8Br_6] \cdot 3CsBr$ or $[Tc_6S_8I_2]$. The electrochemical studies of these complexes in 0.1 M Bu_4NPF_6 in CH_3CN revealed a one-electron reversible reaction of $Tc_6(24e/23e)$ core with the half-waves observed at 0.99 and 0.74 V versus Ag/AgCl for S and Se containing clusters, respectively. These authors reported that the replacements of the cyanide ligands with bromide ones lead to an increase in the $E_{1/2}$ value. This is probably related to π-electron-donating character of the bromides and π-electron-accepting properties of the cyanides. The complexes containing sulfur are characterized by a higher $E_{1/2}$ value as compared with those containing selenium due to differences in the electronegativity of the face-capping chalcogenides. The cyanide complexes of octahetral technetium, $Tc_6(24e)$ or $Tc_6(23e)$, are found to be more thermodynamic stable than their rhenium analogs. This tendency is also observed for axial substituted halide ligands in $[Tc_6S_8X_6]^{4-}$ (M = Re, Tc; X = Br, I) systems. Thiolate capped octahedral hexanuclear technetium(III) clusters were described in next article of Yoshimura et al. (2019). For $[Tc_6(\mu_3 - S)_8Br_6]^{4-}$ the authors observed one–electron reversible reduction at 0.75 V versus Ag, AgCl in 0.1 M $(Bu_4N)PF_6$-CH_3CN assigned to $Tc_6^{III}/Tc^{II}Tc_5^{III}$ couple. Irreversible oxidation of Tc_6^{III} to $Tc^{IV}Tc_5^{III}$ was observed at slightly higher potential of 1.05 V.

Trop et al. (1980b) investigated electrochemical properties of $[Tc(NCS)_6]^{2\,or\,3-}$. These complexes were synthesized as products of a reaction between NH_4NCS and $(NH_4)_2TcX_6$(X = Cl, Br). Figure 3.15 presents the equivalent conductance of the solutions containing various types of technetium complexes dissolved in acetonitrile

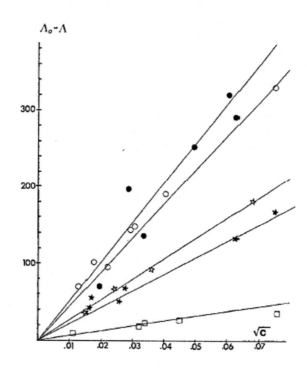

Fig. 3.15 The equivalent conductance of Tc complexes dissolved in acetonitrile: $(n - Bu_4N)(ClO_4)$ □, $(n - Bu_4N)_2[TcCl_6]$ ☆, $(Me_4N)_3[Fe(NCS)_6]$ ○, $(NH_4)_2[Tc(NCS)_6]$ ★, $(n - Bu_4N)_3[Tc(NCS)_6]$ ●. The data are plotted as $\Lambda-\Lambda_0$ — $(cm^2 \cdot \Omega^{-1} \cdot equiv^{-1})$ versus square root of the equivalent concentration (reprinted with permission from Trop et al. (1980b) Copyright 1980 American Chemical Society)

and tetrabutyl perchlorate as supporting electrolytes. Presented data lead to a conclusion that the mobility of $[Tc(NCS)_6]^{3-}$ ions is slightly higher than for $[Fe(NCS)_6]^{3-}$. Further on, $[TcCl_6]^{2-}$ ions are slightly more mobile than $[Tc(NCS)_6]^{2-}$.

Ammonium hexakis(isothiocyanato)technetate(IV) was also extensively studied by Trop et al. (1980b). One-electron reversible $[Tc(NCS)_6]^{2-}/[Tc(NCS)_6]^{3-}$ couple is characterized by a half wave with $E_{1/2} = 0.18$ V versus SCE. Irreversible reduction of $[Tc(NCS)_6]^{3-}$ occurs at $E_{1/2} = -1.09$ V. In turn, oxidation of the $[Tc(NCS)_6]^{2-}$ leads to formation of another one one-electron irreversible wave with $E_{1/2} = 1.60$ V. The value of the standard redox potential of the $[Tc(NO)(NCS)_5]^{2-}/[Tc(NO)(NCS)_5]^{3-}$ couple is equal to 0.138 V versus SCE (see: Clarke and Fackler 1982). The limiting conductance of the electrolytes containing $[NH_4]_2[Tc(NCS)_6]$ (type 2:1) and $[n-Bu_4N]_3[Tc(NCS)_6]$ (type 3:1) in acetonitrile is equal to $\Lambda_0 = 410$ cm$^2 \cdot \Omega^{-1} \cdot$equiv^{-1} and 580 cm$^2 \cdot \Omega^{-1} \cdot$equiv^{-1}, respectively.

The chemistry of technetium and rhenium carbonyls and its halides are of special interest in the context of its medical application as precursors for radiopharmaceuticals. However, discussion of the reactivity of these compounds goes beyond the scope of this monograph and for more information on this topic the reader is referred to the recent review papers (Alberto et al. 2004; Mazzi et al. 2007; Sidorenko et al. 2016).

Colton et al. (1960) analyzed electrochemical properties of rhenium carbonyl and carbonyl halide complexes. Two electrons irreversible waves were observed for all the systems analyzed when absolute ethanol containing 0.3 M methylammonium as a supporting electrolyte was used. Re$^+$-CO system is presumably reduced to the carbonyl rhenide anions. The electroreduction of Re(CO)$_5$I ($E_{1/2} = -1.18$ V) and Re(CO)$_5$Cl ($E_{1/2} = -1.27$ V) is much easier as compared to Re$_2$(CO)$_{10}$ ($E_{1/2} = -1.82$ V). These authors observed that potentials of the half waves of rhenium carbonyl halides decrease in the following order: Cl > Br > I. It can be assumed with caution that the values of the appropriate potentials of technetium carbonyls or technetium carbonyl-halides should be higher than for its rhenium analogs.

In conclusion, despite the huge number of publications that deal with studies on the electroreduction of the pertechnetates and characterization of the reduced technetium species in aqueous and nonaqueous solutions, the understanding of the electrochemical properties of technetium species is still incomplete. As an example, numerous questions concern on the structure and disproportionation of Tc(VI) species present in acidic and alkaline solutions. Further on, most recent reports on the stability of the Tc(VI) are inconsistent with the earlier literature data. At the end of the twentieth century, the unstable Tc(V) was considered as generated exclusively in alkaline solutions. Nowadays, it is well known that the Tc(V), although with a structure different from the species existing at pH > 7, can be generated and stabilized also in strongly acidic media. The mechanism of the Tc(IV) polymerization is also not fully explained. Further on, the Tc(III) containing species await for much deeper exploration and for more detailed electrochemical description. It should be also stressed that the structure of these ions is still unclear as well as the redox chemistry of polymeric Tc(IV) species. Finally, the generation of Tc(II) in aqueous noncomlpexing media is a debatable issue. Therefore, it is very difficult to find reliable values of

Table 3.8 Standard reduction potential of selected technetium red–ox couples (values with uncertainties given are recommended in Rard et al. 1999 and later updates)

Oxidation states of Tc	Redox reaction	$E^{\ominus}/V$
VII/VI	$TcO_4^- + e^- \leftrightarrows TcO_4^{2-}$	-0.64 ± 0.03
VII/V	$TcO_4^- + 2e^- \leftrightarrows TcO_4^{3-}$	-0.60 ± 0.05
VII/IV	$TcO_4^- + 4H^+ + 3e^- \leftrightarrows TcO_2 \cdot xH_2O(s) + (2-x)H_2O$	0.746 ± 0.012
	$2TcO_4^- + 10H^+ + 6e^- \leftrightarrows Tc_2O_2(OH)_2^{2+} + 4H_2O$	0.731 ± 0.011
	$TcO_4^- + 6H^+ + 3e^- \leftrightarrows TcO^{2+} + 3H_2O$	0.660
	$TcO_4^- + 4H^+ + 3e^- \leftrightarrows TcO(OH)_2(aq) + H_2O$	0.595 ± 0.046
VII/III	$TcO_4^- + 8H^+ + 4e^- \leftrightarrows Tc^{3+} + 4H_2O$	0.576
VII/0	$TcO_4^- + 8H^+ + 7e^- \leftrightarrows Tc + 4H_2O$	0.472
IV/III	$TcO(OH)_2(aq) + 4H^+ + e^- \leftrightarrows Tc^{3+} + 3H_2O$	0.516
	$Tc_2O_2(OH)_2^{2+} + 6H^+ + 2e^- \leftrightarrows 2Tc^{3+} + 4H_2O$	0.316
	$TcO_2 + 2H^+ + e^- \leftrightarrows TcO^+ + H_2O$	0.319
III/0	$Tc^{3+} + 3e^- \leftrightarrows Tc$	0.366

numerous thermodynamic and kinetic parameters related to technetium reactions in solutions. Neverethless, the authors of this review made a critical analysis of the literature data reported so far and they select and discuss the most reliable information on Tc chemistry, including the parameters in question. In order to summarize the review, the redox potentials of the selected Tc couples are collected in Table 3.8. This table also includes values not reported by the *OECD Chemical Thermodynamics of technetium*.

References

Alberto R, Pak JK, van Staveren D et al (2004) Mono-, Bi-, or tridentate ligands? The labeling of peptides with 99 mTc-carbonyls. Biopolym (Peptide Science) 76:324–333

Alliot I, Alliot C, Vitorge P et al (2009) Speciation of technetium(IV) in bicarbonate media. Environ Sci Technol 43:9174–9182

Armstrong R, Taube H (1976) Chemistry of trans-Aquontrosyltetraamminetechnetium(I) and related studies. Inorg Chem 15(8):1904–1909

Astheimer L, Schwochau K (1976) Electrochemical reduction of MnO_4^-, TcO_4^- and ReO_4^- in organic solvents: preparation of tetraoxomanganate(VI), -technetate(VI) and -rhenate(VI). J Inorg Nucl Chem 38(6):1131–1134

Astuti Y, Topoglidis E, Gilardi G et al (2004) Cyclic voltammetry and voltabsorptometry studies of redox proteins immobilized on nanocrystalline tin dioxide electrodes. Bioelectrochem 63:55–59

Balasekaran SM, Spandl J, Hagenbach A et al (2014) Fluoridonitrosyl complexes of technetium(I) and technetium(II). Synthesis, characterization, reactions, and DFT calculations. Inorg Chem 53:5117–5128

64   3 Technetium Coordinated by Inorganic Ligands in Aqueous ...

Baldas J, Heath GA, Macgregor SA et al (1998) Spectroelectrochemical and computational studies of tetrachloro and tetrabromo oxo- and nitrido-technetium(V) and their TcVI counterparts. J Chem Soc Dalton Trans 2303–2314
Bancroft EE, Sidwell JS, Blount HN (1981) Derivative linear sweep and derivative cyclic voltabsorptometry. Anal Chem 53:1390–1394
Boyd GE (1978) Osmotic and activity coefficients of aqueous $HTcO_4$ and $HReO_4$ solutions at 25 °C. Inor Chem 17(7):1808–1810
Chatterjee S, Hall GB, Johnsonet IE et al (2018) Surprising formation of quasi-stable Tc(VI) in high ionic strength alkaline media. Inorg Chem Front 5:2081–2091
Chotkowski M, Czerwiński A (2012) Electrochemical and spectroelectrochemical studies of pertechnetate electroreduction in acidic media. Electrochim Acta 76:165–173
Chotkowski M, Czerwiński A (2014a) Thin layer spectroelectrochemical (RVC-OTTLE) studies of pertechnetate reduction in acidic media. J Radioanal Nucl Chem 300:229–234
Chotkowski M, Czerwiński A (2014b) Thin layer spectroelectrochemical studies of pertechnetate reduction on the gold electrodes in acidic media. Electrochim Acta 121:44–48
Chotkowski M, Wrzosek B, Grdeń M (2018a) Intermediate oxidation states of technetium in concentrated sulfuric acid solutions. J Electroanal Chem 814:83–90
Chotkowski M, Grdeń M, Wrzosek B (2018b) Intermediate oxidation states of technetium in alkaline solutions. J Electroanal Chem 829:148–156
Clarke MJ, Fackler PH (1982) The chemistry of technetium toward improved diagnostic agents in: structure and bonding, vol 50. Topics in Inorganic and Physical Chemistry. Springer, p. 62
Cobble JW, Smith WT, Boyd GE (1953) Thermodynamic properties of technetium and rhenium compounds. II. Heats of formation of technetium heptoxide and pertechnic acid, potential of the technetium(IV)-technetium(VII) couple, and a potential diagram for technetium. J Am Chem Soc 75:5777–5782
Colton R, Dalziel J, Griffith WP et al. (1960) Polarographic study of manganese, technetium, and rhenium. J Chem Soc 71–79
Compton RG, Banks CE (2011) Understanding voltammetry, 2nd edn, Ch. 7. Imperial College Press
Cotton FA, Pedersen E (1975) Magnetic and electrochemical properties of transition metal complexes with multiple metal-to-metal bonds. I. $[Tc_2Cl_8]^{n-}$ and $[Re_2Cl_8]^{n-}$ with n = 2 and 3. Inorg Chem 14(2):388–391
Courson O, Le Naour C, David F et al. (1999) Electrochemical behawior of technetium in acetic buffer. Czechoslov J Phys 49(Suppl S1): 687–694
Deutsch E, Heinemann W, Hurst R et al (1978) Production, detection, and characterization of transient hexavalent technetium in aqueous alkaline media by pulse radiolysis and very fast scan cyclic voltammetry. JCS Chem Comm 1038–1040
Duluard S, Ouvrard B, Celik-Cochet A et al (2010) Comparision of PEDOT films obtained via three different routhes through spectroelectrochemistry and the differential cyclic voltabsorptometry method (DCVA). J Phys Chem B 114:7445–7451
Founta A, Aikens DA, Clark HM (1987) Mechanism and kinetics of the stepwise voltammetric reduction of pertechnetates in alkali solutions to Tc(VI), Tc(V) and Tc(IV). J Electroanal Chem 219:221–246
Galus Z (1976) Fundamentals of electrochemical analysis, Chapter 3. Horwood – PWN Warszawa
German K, Kryutchkov SV (2002) Polynuclear technetium halide clusters. Russ J Inorg Chem 47(4):578–583
Gorshkov NI, Miroslavov AE, Lumpov AA et al (2003) Complexation of tricarbonyltechnetium(I) Ion with halide and thiocyanate ions in aqueous solution: 99Tc NMR study. Radiochem 45(2):127–130
Grambow B, Grenthe I, Gaona X et al (2020) Discussion of new data selected for technetium. In: Grenthe I, Gaona X, Plyasunov AV (eds) Second update of the chemical thermodynamics of uranium, neptunium, plutonium, americium and technetium, OECD
Grassi J, Devynck J, Trémilton (1979) Electrochemical studies of technetium at a mercury electrode. Anal Chim Acta 107:47–58

Grassi J, Rogelet P, Devynck J, Trémillon B (1978) Radiopolarography of technetium(VII) n acidic medium. J Electroanal Chem 88:97–103

Guillaumont R, Fanghänel T, Neck V et al (2003) Update on chemical thermodynamics of uranium, neptunium, plutonium, americium and technetium, OECD/NEA. Elsevier p 129

Heineman WR, Hawkridge FM, Blount HN (1984). In: Bard J (ed) Electroanalytical chemistry, vol. 13, Ch. 1. Marcel Deker

Horányi G, Bakos I (1994) Coupled radiometric and electrochemical study of the reduction of TcO_4^- ions at a gold electrode in acidic medium (mechanistic aspects of Tc deposition). J Electroanal Chem 370:213–218

Huber EW, Heineman WR, Deutsch E (1987) Technetium electrochemistry. 5.1 11 spectroelectrochemical studies on the hexachlorotechnetate(IV) and hexabromotechnetate(IV) complexes in aqueous media. Inorg Chem 26(22):3718–3722

Hurst R W, Heineman WR, Deutsch E (1981) Technetium electrochemistry. 1. Spectroelectrochemical studies of halogen, diars, and diphos complexes of technetium in nonaqueous media. Inorg Chem 20(10):3298–3303

Kim B-S, Piao T, Hoier SN et al (1995) In situ spectro-electrochemical studies on the oxidation mechanism of brass. Corr Sci 37(4):557–570

Kissel G, Feldberg SW (1969) Disproportionation of the pertechnetate ion in aqueous alkaline media. An electrochemical study. J Phys Chem 7(9):3082–3087

Könnecke Th, Neck V, Fanghänel Th, Kim JI (1997) Activity coefficients and pitzer parameters in the systems $Na^+/Cs^+/Cl^-/TcO_4^-$ or ClO_4^-/H_2O at 25 °C. J Solution Chem 26(6):561–577

Kryuchkov SV, Pikaev AK, Kuzina AF, Spitsyn VI (1979) Electrolytic dissociation of technetic acid inaqueous solution by pulsed radiolysis. Proc Acad Sci USSR Phys Chem Sect 247:690–692

Kryutchkov S (1999) Chemistry of technetium cluster compounds. Top Curr Chem 176:189–252

Kulesza PJ, Zamponi S, Malik MA et al (1998) Spectroelectrochemical characterization of cobalt hexacyanoferrate films in potassium salt electrolyte. Electrochim Acta 43(8):919–923

Kuzina AF, Zhdanov SI, Spitsyn VI (1962) Polarography of technetium in perchlorate solutions. Dokl Akad Nauk SSSR 144:442–445

Láng GG, Horányi I (2003) Some intresting aspects of the catalytic and electrocatalytic reduction of perchlorate ions. J Electroanal Chem 552:197–211

Lawson BL, Scheifers SM, Pinkerton TC (1984) The electrochemical reduction of pertechnetate at carbon electrodes in aqueous non-complexing acid media. J Electroanal Chem 177:167–181

Li L, Polanco C, Ghahreman A (2016) Fe(III)/Fe(II) reduction-oxidation mechanism and kinetics studies on pyrite surfaces. J Electroanal Chem 774:66–75

Liu X, Poineau F, Fattahi M et al (2005) Speciation of Tc(IV) in chloride solutions by capillary electrophoresis. Radiochim Acta 93:305–309

Lukens WW Jr, Bucher JJ, Edelstein NM et al (2001) Radiolysis of TcO_4^- in alkaline, nitrate solutions: reduction by NO_3^{2-}. J Phys Chem A 105:9611

Magee RJ, Cardwell TJ (1974) Rhenium and technetium. In: Bard AJ (ed) Encyclopedia of electrochemistry of the elements, vol. II. Marcel Dekker, pp. 126–189

Mausolf E, Poineau F, Droessler J et al (2011) Spectroscopic and structural characterization of reduced technetium species in acetate media. J Radioanal Nucl Chem 288(3):723–728

Mazzi U, Schibli R, Pietzsch HJ et al (2007) Technetium in Medicine. In: Zolle I (ed) Technetium-99m pharmaceuticals. Springer, Berlin, Heidelberg

Méndez E, Cerdá MF, Luna AMC et al (2003) Electrochemical behavior of aqueous acid perrhenate-containing solutions on noble metals: critical review and new experimental evidence. J Coll Interfac Sci 263:119–132

Münze R (1968) Redox- und Hydrolysereaktionen von Rhenium- und Technetiumverbindungen. Z Phys Chem 2380:364–376

Nicholson RS (1965) Theory and application of cyclic voltammetry for measurement of electrode reaction kinetics 37(11):1351–1355

Paquette J, Lawren WE (1985) A spectroelectrochemical study of the technetium(IV)/technetium(III) couple in bicarbonate solutions. Can J Chem 63:2369–2373

Pihlar B (1979) Electrochemical behavior of technetium(VII) in acidic medium. J Electroanal Chem 102:351–365

Pitzer KS (1991) Ion interaction approach: theory and data correlation. In: Pitzer KS (ed) Activity coefficients in electrolyte solutions (2nd edn). Chapter 3. CRC Press, pp 75–153

Poineau F, Fattahi M, Montavon et al. (2006) Condensation mechanisms of tetravalent technetium in chloride media. Radiochim Acta 94:291–299

Poineau F, Johnstone EV, Czerwinski KR (2014) Recent advances in technetium halide chemistry. Acc Chem Res 47(2):624–632

Poineau F, Weck PF, Burton-Pye BP et al (2013) Reactivity of HTcO$_4$ with methanol in sulfuric acid: Tc-sulfate complexes revealed by XAFS spectroscopy and first principles calculations. Dalton Trans 42(13):4348–4352

Poineau F, Burton-Pye BP, Sattelberger AP et al (2018) Speciation and reactivity of heptavalent technetium in strong acids. New J Chem 42:7522–7528

Rajec P, Macášek (1981) Electrochemical reduction of Tc(V) chlorocomplexes in acidic medium. J Inorg Nucl Chem 43:1607–1609

Rard JA (1983) Critical review of the chemistry and thermodynamics of technetium and some of its inorganic compounds and aqueous species, Lawrence Livermore National Laboratory. Manuscript

Rard JA, Miller DG (1991) Corrected value of osmotic and activity coefficient of aqueous NaTcO$_4$ and HTcO$_4$ at 25 °C. J Solution Chem 20(12):1139–1147

Rard JA, Rand MH, Anderegg G et al. (1999) Chemical thermodynamics of technetium, vol 3. Elsevier

Rotmanov KV, Maslennikov AG, Zakharova LV, Goncharenko YuD, Pertetrukhin VF (2015) Anodic dissolution of Tc metal in HNO$_3$ solutions. Radiochemistry 57(1):26–30

Rulfs CL, Pacer R, Anderson A (1967) The polarography of aqueous pertechnetate ion. J Electroanal Chem 15:61–66

Russell CD, Cash AG (1978) Polarographic reduction of pertechnetates. J Electroanal Chem 92:85–99

Said KB, Fattahi M, Cl Musikas et al (2000) The speciation of Tc(IV) in chloride solutions. Radiochim Acta 88:567–571

Salaria GBS, Rulfs CF, Elving PJ (1963a) Polarographic behviour of technetium. Anal Chem 2479–2484

Salaria GBS, Rulfs CF, Elving PJ (1963b) Polarographic and coulometric determination of technetium. J Chem Soc 35(8):979–982

Sattelberger A (2005) Technetium compounds. In: Cotton FA, Murillo CA, Walton RA (eds) Multiple bonds between metal atoms, 3rd ed. Springer, pp 251–269

Schwochau K, Astheimer L (1962) Konduktometrische Bestimmung des Diffusionkoefficienten von Pertechnetate-Ionen in wässeriger Lösung. Z Naturforsch 17a: 820–825

Schwochau K, Astheimer L, Hauck J et al (1974) Preparation and properties of tetraoxotechne-tate(VI) and tetraoxorhenate(VI). Angew Chem Internat Edit 13(5):346–347

Sidorenko GV, Maltsev DA, Miroslavov AE et al (2016) Reactivity of higher technetium carbonyls in CO replacement: a quantum chemical analysis. Comp Theor Chem 1093:55–66

Szabó S, Bakos I (2004) Electrodeposition of rhenium species onto a gold surface in sulfuric acid media. J Solid State Electrochem 8:190–194

Trop HS (1979) Synthesis, characterization and reactivity of technetium and rhenium complexes in intermediate oxidation states. Ph.D. thesis, Massachusetts Institute of Technology

Trop HS, Jones AG, Davison A (1980a) Technetium cyanide chemistry: synthesis and characteri-zation of technetium(III) and (V) cyanide complexes. Inorg Chem 19:1993–1997

Trop HS, Davison A, Jones AG et al. (1980b) Synthesis and physical properties of hexakis(isothiocyanato)technetate(III) and -(IV) complexes. Structure of the [Tc(NCS)$_6$]$^{3-}$ Ion. Inorg Chem 19(5):1105–1110

Vichot (2001) Spéciation du technétium en milieu chloro-sulfaté. Contribution à l'etude des effects de la radiolyse γ. Ph.D. thesis, Universite Paris XI UFR Scientifique D'Orsay

Vichot L, Fattahi M, Musikas C et al (2003) Tc(IV) chemistry in mixed chloride/sulphate acidic media. Formation of polyoxopolymetallic species. Radiochim Acta 91:263–271

Vongsouthi N (2009) Spéciation du technétium-99 en milieu acide noncomplexant: effet Eh-pH. Ph.D. thesis, Université de Nantes, UFR Sciences et Techniques

Wang Y-H, He J-B (2012) Corrosion inhibition of copper by sodium phytate in NaOH solution: cyclic voltabsorptometry for in situ monitoring of soluble corrosion products. Electrochim Acta 66:45–51

Yalçıntaş E, Gaona X, Altmaier M et al (2016) Thermodynamic description of Tc(IV) solubility and hydrolysis in dilute to concentrated NaCl, MgCl$_2$ and CaCl$_2$ solutions. Dalton Trans 45:8916–8936

Yoshimura T, Ikai T, Takayama T et al (2010) Synthesis, spectroscopic and electrochemical properties, and electronic structures of octahedral hexatechnetium(III) clusters [Tc$_6$Q$_8$(CN)$_6$]$^{4-}$ (Q = S, Se). Inorg Chem 49:5876–5882

Yoshimura T, Nagata K, Matsuda A et al (2019) Synthesis, structures, redox properties, and theoretical calculations of thiohalide capped octahedral hexanuclear technetium(III) clusters. Dalton Trans 48:14085–14095

Zamponi S, Czerwinski A, Marassi R (1989) Thin-layer derivative cyclic voltabsorptometry. J Electroanal Chem 226:37–46

Chapter 4
Technetium Coordinated by Organic Ligands in Aqueous and Nonaqueous Solutions

Technetium organic complexes are explored mainly in the context of their application in nuclear medicine as imaging agents. It is known that drugs biodistribution strongly depends on the charge of the species. As an example, reduction of Tc species by, e.g., glutathione or thioredoxin proteins to neutral or opposite charged species may result in unwanted loss of radiopharmaceutical efficiency and the target specificity. The electrochemical studies on Tc complexes may provide information very helpful in prediction of their behavior in, e.g., fluids present in a human body. As an example, the determination of the redox potentials of selected Tc forms is important for evaluation of their suitability for medical applications. However, in contrast to inorganic technetium compounds only a handful of papers deal with studies on electrochemical properties of technetium ions coordinated by organic ligands in aqueous and nonaqueous solutions. Further on, these papers usually report the results of electrochemical or spectroelectrochemical measurements only as an addition to the results of other methods applied as the main ones. The cycle of works by Deutsch, Heinemann and coworkers, which focus exclusively on electrochemical studies on phosphino-Tc(III) complexes is an exception from the rule (e.g., Hurst et al. 1981; Ichimura et al. 1984, 1985; Kirchhoff et al. 1987, 1988; Konno et al. 1988; Wilcox and Deutsch 1991). Other examples of papers that are focused on electrochemical studies of discussed Tc species include works by Refosco et al. (e.g. 1988) who analyzed oxo-technetium(V) complexes containing Schiff bases ligands and works by Pinkerton and Heineman whose paper deals with an analysis of the electroreduction of pertechnetates in phosphonate media (Pinkerton and Heineman 1983).

Cyclic voltammetry is the technique most common applied in studies on Tc compounds stability. It allows determining the half-wave potential, $E_{1/2}$, of the considered redox system. The redox properties of the technetium organic complexes discussed later in this Chapter are briefly summarized in Table 4.3.

© Springer Nature Switzerland AG 2021
M. Chotkowski and A. Czerwiński, *Electrochemistry of Technetium*,
Monographs in Electrochemistry, https://doi.org/10.1007/978-3-030-62863-5_4

4.1   Nonaqueous Solutions

Halogen (X), diars or diphos (D) complexes of technetium ($[Tc^{III}D_2X_2]^+$) in DMF were analyzed by Hurst et al. (1981). These species undergo a reversible one-electron redox reaction with participation of Tc(III)/Tc(II) couples, Eq. (4.1):

$$[TcD_2X_2]^+ + e^- \rightleftharpoons [TcD_2X_2]^+ \tag{4.1}$$

Additional redox reactions of Tc diars complexes include Tc(II)/Tc(0) although description of these processes is less complete than for Tc(III)/Tc(II) systems. The authors noted that the reduction potential of Tc(III) is sensitive to the nature of both the halogen and the organic ligand. Thus, replacement of a light halogen with a heavier one leads to an increase in the reduction potential by 0.07–0.12 V. It was proposed that this effect is a result of π-back-bonding and stabilization of d^5 (Tc(II)) over d^4 (Tc(III)) configuration by increased acceptance of t_{2g} electron density. The diphos ligand stabilizes d^5 (Tc(II)) over d^4 (Tc(III)) center more effective than the diars ligand.

Later work of Heineman, Deutsch et al. (Libson et al. 1983; Jurisson et al. 1984; Bandoli et al. 1984; Ichimura et al. 1984) were devoted also to technetium(III) complexes of $[TcD_2X_2]^+$ type where D stands for (dppb), (dppe), (dmpe) or (depe). CV curves reveal one-electron reversible reduction reactions of Tc^{III} to Tc^{II} and Tc^{II} to Tc^I and a one-electron oxidation of Tc^{III} to Tc^{II} (Jurisson et al. 1984; Ichimura et al. 1985).

Mazzi et al. (1980) examined Tc(I) complexes with phosphine and CO ligands by means of the voltammetric methods. [$TcCl(CO)_2(PMe_2Ph)_3$] and [$TcCl(CO)_3(PMe_2Ph)_2$] undergo a two-electron process, which leads to the formation of Tc(III). Electrooxidation of Tc(I) in $TcCl(CO)_2(PMe_2Ph)_3$ results in the formation of anodic waves seen on the voltammetric curves at 0.83 V (peak A) and 1.02 V (peak B) (Fig. 4.1). The first reduction peak C at 0.74 V is linked with the anodic peak A. ΔE_p of this redox couple equals 0.09 V and its electrochemical behavior cannot be considered as a reversible one under diffusional control. A small peak at -0.65 V is attributed to an irreversible reduction process. The authors noted that the position and height of the second oxidation peak (B) vary and depend on the electrode history as a consequence of adsorption of the technetium species. The shape of the CV curve for the second-investigated Tc(I) complex, [$TcCl(CO)_3(PMe_2Ph)_3$], was similar to that recorded for $TcCl(CO)_2(PMe_2Ph)_3$ although for the former the anodic peaks are observed at more positive potentials (1.17 and 1.54 V). The cathodic peak of $TcCl(CO)_2(PMe_2Ph)_3$ reduction was observed at 1.08 V indicating the same as previously peak separation (0.09 V). The conductivity of 1 mM [$TcCl(CO)_2(PMe_2Ph)_3$] in ACN was determined at a level of 230 Ω^{-1} cm^2 mol^{-1}.

Figure 4.2 presents a complex scheme of chemical and electrochemical steps of Tc complexes oxidation. Noteworthy is the fact that the Tc(II) species may undergo disproportionation or chemical addition of ACN.

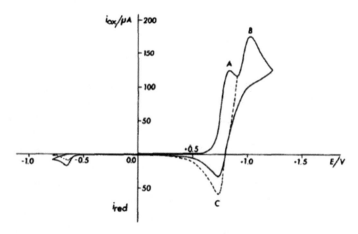

Fig. 4.1 Cyclic voltammetric curves recorded on a 3.1 mM TcCl(CO)$_2$(PMe$_2$Ph)$_3$, 0.1M NaClO$_4$, ACN sol.; potential scan initially anodic in direction. Pt working microelectrode. Scan rate 0.2 V s^{-1} (E given vs. SCE) (reprinted with permission from Mazzi et al. 1980 Copyright 1980 Elsevier)

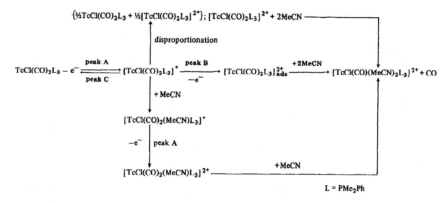

Fig. 4.2 Scheme of [TcCl(CO)$_2$(PMe$_2$Ph)$_3$] transformation (reprinted with permission from Mullen et al. (2000) Copyright 1980 Elsevier)

A later paper by Wang et al. (1993) reports results of studies on complexes of Tc(I) with trans-[TcH(L)(dppe)$_2$] where (L = N$_2$, CO or CNR). The authors described an irreversible one-electron oxidation process of carbonyl and dinitrogen technetium(I) complexes. A substitution of (CN-tert-buthyl) or (CNC$_6$H$_{11}$) for (CO) or (N$_2$) resulted in a quasireversible oxidation of Tc(I) to Tc(II). The $E_{p/2}^{ox}$ of the complexes studied increases in the order of CNR < N$_2$ < CO and this effects can be explained by differences in donor abilities of the ligands. Due to a limited number of papers devoted to the analysis of the carbonyl Tc compounds, it is worth to recall the results of similar studies carried out with Tc analogs, i.e., Mn and Re. Thus, Bond et al. (1978) discussed the electrochemical properties of Mn(II) and Re(II) tricarbonyl derivatives (mer-[M(CO)$_3$(phosphine)$_2$X]). They concluded that the examined M(II) complexes are strong oxidants ($E^{0'}$ M(II)/M(I) > 1.15 V in DCM vs. Ag, AgCl). One may expect the same behavior also for Tc(II) complexes.

Linder et al. (1986) synthesized nitrosyl complexes of Tc(I) and reported results of electrochemical studies of these species. An irreversible reduction of $[Tc(NO)(CNCMe_3)_5]^+$ leads to the formation of a single reduction peak at − 0.72 V (vs. SCE). This complex turned out to be stable even in the presence of water. A quasireversible oxidation of Tc(I) bonded in $[Tc(CNCMe_3)_6]$ to Tc(II) was observed at 0.82 V versus SCE. Synthesis and preliminary electrochemical studies of $[Tc^{II}(NO)(CF_3COO)_4F]^{2-}$ complex were recently reported by Balasekaran et al. (2017). CV curves recorded for this complex in trifluoroacetic acid show a one-electron reduction wave at −0.744 V versus Ag, AgCl (3M KCl). This value was close within ±0.1 V to those measured for other inorganic nitrosyl-technetium complexes and reported earlier by Armstrong and Taube (1976) or Balasekaran et al. (2014). The separation of anodic and cathodic peak was equal to 0.125 V, which excludes reversibility of the process.

$[TcD_3]^+$, which electrochemistry was studied by Ichimura et al. (1984), was found to be relatively stable in aqueous solutions. The oxidation of $[Tc(depe)_3]^+$ takes place at a potential lower by 0.164 V than the respective reaction of $[Tc(dmpe)_3]^+$. This observation indicates that the dmpe ligand is a less effective σ donor than depe. The $E^{0'}$ for Tc(II)/Tc(I) couples in $[TcD_3]^{2+/+}$ complexes turned out to be 1.57–1.83 V higher than the respective values for the corresponding $[Tc^{II}D_2X_2]^{0/-1}$ complexes. It means that stabilization of spin-paired d^6 (Tc(I)) is stronger for bis(phosphinine) than for two halide ligands. Reduction of $[Tc^{III}D_2X_2]^+$ in 0.5M KNO$_3$ leads to the formation of water insoluble $[Tc^{II}D_2X_2]^0$. The complexes of this type were discussed in an early work of Kirchhoff et al. (1988). $E^{0'}$ of more water soluble Tc(II)-dmpe determined in aqueous solutions was found to be higher by ca. 0.2 V than the value measured in nonaqueous media (Ichimura et al. 1984). Additional experiments carried out in various solvents (DMF, propylene carbonate, acetonitrile) have shown that stabilization of dipositive charged Tc(II) complexes increases when the solvent becomes more basic. Later work of Ichimura et al. (1985) deals with an analysis of influence of properties of selected Schiff bases and monodentate tertiary phosphine ligands on the electrochemistry of technetium(III) complexes in propylene carbonate solutions. A one-electron reversible reduction of Tc(III) to Tc(II) and a one-electron reversible oxidation of Tc(III) to Tc(IV) were observed for trans-$[Tc^{III}(PR_2R')_2L]^+$ where PR$_2$R' is a monodentate tertiary phosphine with R, R' indicating ethyl/phenyl and L denoting a tetradentate Schiff base ligand. The difference in $E^{0'}$ values between Tc(III)/Tc(II) and Tc(IV)/Tc(III) couples turned out to be in the range from 1.5 V to 1.75 V indicating high stability of the examined Tc(III) complexes.

Complexes of technetium with sexidentate Schiff-base ligands were investigated by Hunter and Kilcullen (1989). A one-electron reduction of $[Tc^{IV}L^{3'}]^+$ to neutral $[Tc^{III}L^{3'}]$ was observed and the authors suggest accumulation of such formed Tc(III) species at the electrode surface. It is not clear whether the process is reversible or not. Later, Refosco et al. (1993) synthesized neutral complexes of Tc(III) with O,P-bidentate phosphino-carboxylate ligands ($[Tc(L^n)_3]$, $n = 1, 2, 3$, see abbreviation list for Table 4.2 for details). All the investigated systems revealed the existence of Tc(III)/Tc(II) redox couples and quasi-reversible one-electron Tc(IV)/Tc(III) systems. Although ligands with higher acidity should stronger stabilize the complex due to favored delocalization of the charge density from the Tc(II), the reduction of $[Tc(L^1)_3]$ complex with the strongest π-acid ligand (phosphino-benzoic) takes place at potentials lower than for the other complexes. In this case, steric factors are so important that they change redox properties of the couple significantly.

In 1991, Wilcox with Deutsch (1991) published an article that was a continuation of earlier works on Tc complexes with derivatives of phosphines as ligands. The research focuses on the redox properties of Tc(III/II) complexes containing dimethyl(or diethyl)phenylphosphines and polypyridyl as ligands. Also this time, the reactions of Tc(IV)/Tc(III) and Tc(III)/Tc(II) redox couples containing (bypy) turned out to be diffusion controlled one-electron processes. (Me_2bpy) stabilizes higher oxidation states of Tc, e.g., Tc(IV) more effectively than (bpy) ligands because the former ligand is a better σ donor and poorer π acceptor than the latter. Moreover, Tc(IV) complexes are generally less stable than corresponding Tc(II) containing species. Early work of Breikss et al. (1990) deals with electrochemical studies on properties of Tc(III) coordinated by mixed phosphine, chloride and nitrogen-donors. Also for these systems, a reversible Tc(IV)/Tc(III) couple was observed at potentials close to those reported by Wilcox and Deutsch (1991).

An increase in the number of phenyl groups in phosphine ligands facilitates the electroreduction of Tc(III) to Tc(II) but impedes electrooxidation of Tc(III) to Tc(IV). The ligands with Ph- groups are stronger π-acids and more efficiently stabilize Tc(II) in opposite to alkyl-substituted ligands, which are more effective σ-donors and better stabilize Tc(IV). The halogens are the other ligands able to stabilize the lower oxidation states of Tc and this effect is somewhat stronger for the heavier halogens.

Konno et al. (1988, 1989a, b, 1992a, b, 1993) studied electrochemistry of Tc complexes containing phosphines and thiolato ligands. The results indicate a one-electron reversible reduction of Tc(III) to Tc(II). The redox properties of Tc(II)/Tc(I) and Tc(IV)/Tc(III) strongly depend on the ligand properties. For example (depe), which is a σ-donor better than (dmpe), provides a stronger stabilization of the higher Tc oxidation states. Moreover, replacement of the halogens with stronger σ-donating thiolato ligands in Tc(III) complexes shifts the $E^{0'}$ values by several hundred mV toward more negative ones. Aromatic substituents in such types of ligands enhance their π-acid properties and this results in $E^{0'}$ values more positive than for alkyl substitutes. The $E^{0'}$ of examined Tc-complexes decreases according to changes of thiolato ligands structures in the following order: SC_6H_4-p-Cl > SC_6H_5 > SC_6H_4-p-CH_3 > $SCH_2C_6H_5$ > $SCH_2C_6H_4$-p-OCH_3 > SC_2H_5. An increase in the length of the alkyl group in the ligand makes the Tc(III) complex more difficult to reduce but easier to oxidize.

Konno et al. (e.g. 1989b) discussed the influence of the spatial distribution of the ligands in a complex on the redox properties of the latter. The steric repulsion between various groups as well as ligand competition for the electron density of Tc center are among the most important factors that determine the stability of Tc complexes. For example, cis-geometry in $[Tc(SC_6H_5)_2(diars)_2]$ makes its reduction more difficult (by 23 mV) in comparison to cis-[Tc(SC$_6$H$_5$)$_2$(dmpe)$_2$] due to the rigidity of (diars) ligand and its lower ability to function as a π-acid.

Similar technetium complexes were examined also by Dilworth et al. (1992). Tc(III or V) were coordinated by 2-(diphenylphosphino)benzenethiol as the ligand. Neutral [Tc^{III}(dppbt)$_3$] exhibited reversible redox processes, i.e., two oxidation reactions and one reduction process. This is in contrast to [Tc^{V}OCl(dppbt)$_2$], which does not undergo oxidation processes even at potentials as high as 1 V. Moreover, reduction of the latter complex becomes reversible only at extremely low temperatures.

Crown thioether (trithiacyclononane) as a ligand in Tc-complexes was described by White et al. (1992) and Mullen et al. (2000). They observed reversible or quasi-reversible redox Tc(III)Tc(II) couples. Pasqualini and Duatti (1992) reported a high stability of [TcN(noet)$_2$] complex where (noet) is N-ethyl-N-ethoxydithiocarbamate ligand. These species do not reveal any oxidation or reduction signals in ACN solutions in a wide potential window from -1.75 V to c.a. 1.225 V versus SCE. For the latter potential, an irreversible two-electron oxidation process was observed.

Technetium complexes with mixed phosphine and dithiocarbamate ligands were of interest to Okamoto et al. (1993a). All the examined systems containing dithiocarbamate ligands were characterized by reversible Tc(III)/Tc(II) and Tc(II)/Tc(I) couples observed at ca. 0.3 and -0.5 V versus Ag, AgCl (3M NaCl), respectively. The trans-geometry is preferred for [Tc^{III}(SR)$_2$(dmpe)$_2$]$^+$ (R-alkyl or benzyl) complexes with saturated carbon bounded to sulfur while for phenyl or phenyl derivatives acting as R the cis-geometry is usually observed. The redox potentials of technetium coordinated by zwitter-ionic ligand (SCP) and phosphine are shifted toward more negative potentials as compared with other discussed systems. Apart from reversible Tc(III)/Tc(II) couple also irreversible Tc(II)/Tc(I) system was observed (Okamoto et al. 1993b).

Technetium complexes with tridentate dithiolato fragments were of interest to Pietzsch et al. (2001, 2003). According to these authors, these types of nonpolar, liphophilic and sterically shielded oxo-free Tc(III) complexes can be used for the synthesis of new radiotracers. The redox stability window of synthesized "3 + 2" complexes turned out to be equal to ca. 0.7–0.8 V being ca. 0.1–0.2 V larger than for Tc(III) "3 + 1 + 1" species. "Sulfur rich" bis(perthiobenzoato)(dithiobenzoato)technetium(III) was of interest to Mévellec et al. (2002). Two one-electron reversible Tc(IV)/Tc(III) and Tc(III)/Tc(II) couples were observed for this complex. These redox couples were separated by 1.45 V. It turned out that the examined Tc complex is more difficult to oxidize and easier to reduce than its Re analog.

Technetium complexes with chelating thiourea ligands, including arylthioureas and N-(N'', N''-dialkylaminothiocarbonyl)-N'-substituted benzamidines, were of interest to Huy (2009). This author noted that Tc(III) with tridentate benzamidine ligands, [Tc^{III}(PPh$_3$)(L^{1b})(morphbtu)] can be oxidized in two well-separated (0.86 V) one-electron steps to [Tc^V(PPh$_3$)(L^{1b})(morphbtu)]$^{2+}$. The latter form turned out to be stable in the presence of Ar. Oxo-Tc(V) complex was described as a more thermodynamically stable in the presence of air than the Tc(III) complex.

Tc^{V}O-triarylcorroles were of interest to Einrem et al. (2016). They observed that the Tc, which is the coordination center of the complex, is inactive in redox reactions. The electrochemical HOMO–LUMO gap was equal to 2.04–2.09 V, which represented the difference in π and π^* energies of the macrocycle.

Tc(III) complexes with tetradentate N_2O_2 Schiff base are intensively investigated today as potential multidrug-resistant tumor or myocardial perfusion imaging agents (Baumeister et al. 2018). Reported [Tc(tmpp)$_2$(tmf$_2$en)]$^+$ and [Tc(PEt$_3$)$_2$(tmf$_2$en)]$^+$ undergo a one-electron reversible reduction at potentials about 0.2–0.23 V higher than their rhenium analogs. The oxidation of such types of Tc(III) complexes is irreversible and occurs at ca. 0.82 V versus Ag, AgCl.

Pyridine complexes of technetium which have +3, +2 and +1 valencies were analyzed by Barrera et al. (1996). At least two (IV/III and III/II or III/II and II/I) reversible couples were observed for the examined systems. The $E^{0'}$ of Tc(III/II) becomes increasingly more negative in the following order of the amine ligands: tpy < Me$_2$bpy < py < tmeda. It suggests that polypyridyl ligands stabilize more stronger Tc(II), relative to Tc(III), than pyridyl ligands. A substitution of chlorides with a pyridine ligand results in a shift of $E^{0'}$ of Tc(II)/Tc(I) couple by 0.39 V. Nicholson et al. (1991) examined tris-diazene chelate complexes of Tc(I). Quasi-reversible processes were observed for [Tc(C$_8$H$_5$N$_2$N = NH$_3$)$_3$]$^+$ at potentials of -0.24 V and 0.004 V versus Ag, AgCl at room temperature.

A very complex electrochemical behavior of boron-capped technetium-dioxime complexes was reported by Cyr et al. (1993). The species with the formula of TcX(dioxime)$_3$BR (where X = Cl, Br or OH and R = OH, Me, Et, Bu or Ph) were examined using CV, d.c. polarography and coulometry. The CV curves revealed the existence of three characteristic redox features. The authors reported that chloro- and bromo-complexes undergo a two-electron irreversible reduction in ACN with possible decomposition at ca. -1.3 V versus Ag/Ag$^+$. The products of this reaction are also irreversibly oxidized at c.a. -0.7 V. The second two-electron reduction peak was observed at c.a. -2.7 V. The authors concluded that both reduction and oxidation of boron-capped Tc-dioxime complexes appeared to be biologically inaccessible.

Patterson et al. (1986) studied electrochemistry of Tris(β-diketonato)technetium(III) and (IV). The redox properties of Tc(IV)/Tc(III) and Tc(III)/Tc(II) redox couple strongly depend on the nature of the ligand. Tris(dipivaloylmethanato)technetium(III) was reversible oxidized at 0.12 V; tris(trifluoroacetylacetonato)technetium(III) was characterized by irreversible current waves at 1.0 and -1.16 V while tris(hexafluoroacetylacetonato)technetium(III) was quasi-reversible reduced at 0.14 V versus SCE in ACN.

The experiments carried out by Luo (1995) were focused on synthesis and analysis of chemical properties of new hexadentate ligand. Cyclic voltammetry curves recorded for [Tc^{III}(HP$_2$O$_4$)] where H$_4$P$_2$O$_4$ is P,P,P',P'-tetrakis (o-hydroxyphenyl)diphosphinoethane at potentials from -2 to 2 V versus Ag, AgCl and at a scan rate of 0.1 V·s^{-1} do not reveal electrochemical signals between -2 and 2 V versus Ag, AgCl indicating a very high stability of this complex.

Much more complex electrochemical behavior of technetium species was observed for oxo-Tc(V) species containing Schiff bases (Refosco et al. 1988; Tisato et al. 1989, 1990). A loss of a ligand in the cis- position to Tc = O bond was discussed as a result of a one-electron reduction of Tc(V) (in TcOCl(L$_B$)$_2$) to Tc(IV). Electrogenerated Tc(IV) complex undergoes isomerization to more stable product with vacant site trans- to Tc = O linkage. The latter complex is reduced to Tc(III), which also undergoes isomerization. Based on the results, the authors stated that 8-quinolinol ligands (especially 5-nitrido derivatives) effectively stabilize Tc(IV) and Tc(III) oxidation states. Tisato et al. (1989) observed two well-developed one-electron redox couples for quinquedentate Schiff base ligands. Linden et al. (1994) studied electrochemistry of technetium(V)oxo nitroimidazole complex, which is a very promising imaging agent that can be applied in the analysis of regional hypoxia. The CV curve has shown an irreversible process at low potentials.

Bolzati et al. (1997) successfully synthesized nitrido Tc(V) complexes with ferrocenedithiocarboxylate. Electrochemical studies of this type of complexes reveal the existence of quasi-reversible couples located very close to Fe(III)-Fe(II) systems characteristic for free FcCS$_2$ ligand.

Unusual Tc(VI) complexes with 3,5-di-tert-butylcatechol (DBCat) ligand were synthesized by deLearie et al. (1989). Tc(DBCat)$_3$ was synthesized in a simple reaction between pertechnetates and the ligand. The latter played the dual role of reductor and complexing agent. CV curves recorded for these complexes revealed three one-electron reversible reactions, including Tc(VI)/Tc(V) and Tc(V)/Tc(IV) couples and reversible reactions of the ligand. The redox process characteristic for this ligand was shifted toward more positive values by 0.86–0.87 V as compared with Tc(VI)/Tc(V). Additional two irreversible oxidation processes took place at much higher potentials (c.a 1 V and 1.3 V vs. Tc(VI)/Tc(V) couple). A comparison of the electrochemical properties of the respective Tc and Re complexes revealed that cationic technetium complex is more stable than its rhenium analogs.

Technetium of a low valency (especially +4, +3) can also exist as μ-oxo technetium complexes. For example, Kastner et al. (1986) or Clarke et al. (1988) examined such types of complexes with halide and pyridine ligands. CV curves of these compounds revealed two well-separated one-electron redox processes. The dissymmetric ligands (1. type: $[X(L)_3XTc\text{-}O\text{-}TcX(L)X_3]$) turned out to be generally less stable than the asymmetric ones (2. type: $[X(L)_4Tc\text{-}O\text{-}Tc(L)X_4]$). The differences between $E_{1/2}$ of Tc^{IV}-Tc^{III}/Tc^{III}-Tc^{III} and Tc^{III}-Tc^{III}/Tc^{III}-Tc^{II} redox couples were equal to 0.97–1.16 V and 1.35–1.63 V for the first and the second type of Tc complexes, respectively. Later, Linden et al. (1989) synthesized a dimeric Tc^{IV}-Tc^{IV} complex, $[(\text{TCTA})Tc(\mu\text{-O})_2Tc(\text{TCTA})]^{2-}$. Voltammetric curves recorded for these species revealed the existence of three redox systems therein pH independent Tc(IV-IV)/Tc(IV-III) couple at a potential of 0.167 V versus SCE.

New oxo-bridged complexes of Tc containing bipyridine were described in the next article by Lu et al. (1993). Three well-separated one-electron redox couples were observed. Noteworthy is the fact that $(\mu\text{-O})[X(L)_2Tc]_2^{2+}$ was relatively stable

Table 4.1 Conductivity of selected Tc-phenylphosphonite complexes (Biagini Cingi et al. 1975; Mazzi et al. 1977)

Compound	$\Lambda_{eq}/\Omega\ mol^{-1}\ cm^2$
[TcCl$_2$L$_4$]Cl	80.3
[TcCl$_2$L$_4$]BPh$_4$	76.1
[TcCl$_2$L$_4$]ClO$_4$	75.9
[TcBr$_2$L$_4$]ClO$_4$	70.4
[TcI$_2$L$_4$]ClO$_4$	71.6
[TcI$_2$L$_4$]I	68.5
trans-[Tc(CO)$_2$L$_4$]ClO$_4$	93.8
cis-[Tc(CO)$_2$L$_4$]ClO$_4$	80.2

even in aqueous solutions. Moreover, the $E^{0'}$ measured in aqueous solutions for Tc^{IV-III}/Tc$^{III-III}$ redox couples remained constant for all the complexes examined.

Barrera and Bryan (1996a) examined electrochemically oxo-bridged technetium(III) polypyridyl complex. Based on recorded CV curves, they concluded that the metal–metal bonds are relatively weak in this compound. The electroreduction of Tc^{III}-Tc^{III} to Tc^{III}-Tc^{II} as well as Tc^{III}-Tc^{II} to Tc^{II}-Tc^{II} turned out to be one electron reversible processes separated by 0.25 V.

Metal nitrile species with metal–metal bonding were the topic of a cycle of papers published in 1990s (see e.g., Bryan et al. 1995; Cotton et al. 1996a, b, 1997). Bryan et al. (1995; Cotton et al. 1997) examined [Tc$_2$(CH$_3$CN)$_{10}$]$^{4+}$ as an example of such type of the complexes. Deeper reduction of this dinuclear complex to Tc^{II}-Tc^{I}/Tc^{I}-Tc^{I} weakens the metal–metal bond causing an irreversible process. They also observed that [TcCl$_2$(CH$_3$CN)$_4$]$^+$ complex undergoes the reduction easier than [Tc$_2$Cl$_2$(py)$_4$]$^+$. Cotton et al. (1996a), in turn, described phenyl-phosphine complexes of [Tc$_2$Cl$_5$(PMe$_2$Ph)$_3$] and [Tc$_2$Cl$_4$(PMe$_2$Ph)$_4$] types with a Tc–Tc dinuclear core. Replacement of the Cl$^-$ ligand with a phosphine resulted in a shift of $E^{0'}$ toward more negative values by 0.39 V and 0.48 V for the oxidation and the reduction process, respectively. (Tc^{III}-Tc^{II})-formamidinate (DPhF) complexes with $\sigma^2\pi^4\delta^2\delta^*$ ground state configuration were described in a next publication of this group (Cotton 1996b). Noteworthy is the fact that the same configuration is reported also for inorganic Tc–Tc dinuclear complexes, [Tc$_2$Cl$_8$]$^{3-}$ with a formal bond order of 3.5 (Jones and Davison 1982). Replacement of two chloride atoms with a formamidinate ligand makes the higher oxidations states of Tc more stable due to its greater π-basicity. In general, the scheme of the redox properties of the dinuclear technetium core is complex, as shown in Fig. 4.3.

Dinuclaer technetium complexes were of interest also to Poineau et al. (2010). A comparison of Tc$_2$Cl$_4$(PMe$_3$)$_4$ and Tc$_2$Br$_4$(PMe$_3$)$_4$ revealed an inverse halide order effect (IHO) based on a back bonding effect. The bromide Tc-complex turned out to have a higher oxidation potential than the chloride Tc-complex.

Just a handful of papers deal with studies on transport properties of the Tc complexes (e.g. Biagni Cingi et al. 1975; Mazzi et al. 1977). These authors determined the conductivity, Λ_{eq}, of 1 mM solutions of hexacoordinated technetium(III)

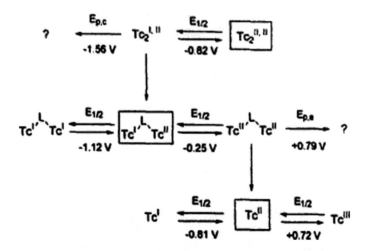

Fig. 4.3 Electrochemical relationships based on cyclic voltammetry between the various homoleptic acetonitrile complexes of technetium. Redox potentials are given as volts versus Cp_2Fe, Cp_2Fe^+ in acetonitrile. Species enclosed by rectangles have been chemically isolated and identified. $Tc^n = [Tc(CH_3CN)_6]^{n+}$, $Tc^{n,m} = [Tc(CH_3CN)_6]^{(n+m)+}$. Tc^n-L-$Tc^m = [Tc_2(\mu,\eta^1,\eta^2\text{-}(CH_3CN)(CH_3CN)_{10}]^{(n+m)+}$ (reprinted with permission from Cotton et al. (1997). Copyright 1997 Elsevier)

complexes with diethylphenylphosphonite ($L = P(OEt)_2Ph$) in nitromethane at 25 °C. The respective values are listed in Table 4.1. Additionally, these authors reported that $[TcX_2L_4]$ complexes turned out to be stable in polar solvents (e.g., ethanol, acetone) in the presence of free ligands.

4.2   Aqueous Solutions

The electrochemical properties of Tc organic complexes in aqueous solutions are less often studied than for nonaqueous solvents. Works of Russell and Speiser (1982) are examples of such studies. These authors investigated the electroreduction of pertechnetates in the presence of iminodiacetates (DPTA: diethylenetetraaminepentaacetate; EDTA: dethylenediaminetetracetate; ADA: N-(2-acetamido)iminodiacetate or HIDA: N-(2,6-dimethylphenylcarbomoylmethyl)iminodiaeetate) over a broad range of pH. The influence of pH on the polarographic half-wave of TcO_4^- ions electrodeduction in the presence of DPTA is presented in Fig. 4.4.

Russell and Speiser concluded that at pH lower than 6, the pertechnetates are quantitatively reduced to Tc(III) species. This is in contrast to their behavior in solutions with higher pH values where generation of Tc(IV) species is observed instead. In neutral solutions, the product of the electroreduction was identified as a mixture of Tc(III) and Tc(IV) species. The reoxodation of Tc(III) leads to the formation of Tc(V).

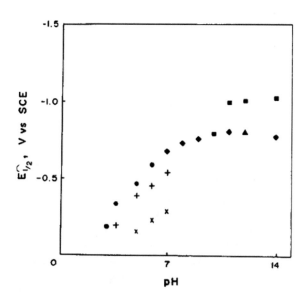

Fig. 4.4 Polarographic half-wave potentials versus pH in 0.1M EDTA. ● Tc(VII) → Tc(III): ■ Tc(VII) → Tc(IV): ▲Tc(VII) → Tc(V); ◆ Tc(VII) → mixed oxidation states: + Tc(III) → Tc(IV)?; × Tc(III) → Tc(V)? (reprinted with permission from Russell and Speiser (1982). Copyright 1982 Elsevier)

Noteworthy is the article published by Pinkerton and Heineman (1983). These authors not only summarized earlier works on the electroreduction of pertechnetates in the presence of phosphate ligands but also discussed the results of studies on the reduction of pertechnetates in aqueous hydroxyethylidene diphosphate (HEDP) media. This process was examined at pH of 3, 5 and 7 using a hanging mercury drop electrode. An initial electroreduction step revealed to be an irreversible process with a heterogeneous electron transfer rate constant strongly dependent on pH, Table 4.2.

In neutral and slightly acidic solutions, the discussed process leads to the formation of Tc(V) species and was found to be a second-order reaction in respect to hydrogen ion concentration. At pH of 3, the process in question leads to the formation of Tc(IV). The latter conclusion is in line with results of Pihlar who also reported that heterogeneous rate constant for the electroreduction of pertechnetates in 0.15M NaCl at pH 1.8–3.1 is equal $2.4 \cdot 10^{-4}$ cm s^{-1} (Pihlar 1979).

Pinkerton and Heineman concluded that the electroreduction of pertechnetates in the presence of a complexing agent could follow an ECEC mechanism according to Eqs. (4.2)–(4.5):

Table 4.2 Heterogeneous electron transfer rate constants for primary reduction of pertechnetates in buffered HEDP and charge transfer of this process (Pinkerton and Heineman 1983)

pH	$\alpha n \pm$ S.D.	k_{fh}^0/cms^{-1}	$n \pm$ S.D.	$E^{0'}$/V
7.4	0.65 ± 0.05	$1.0 \cdot 10^{-8}$	1.87 ± 0.17	-0.624
5.0	0.58 ± 0.09	$4.9 \cdot 10^{-6}$	2.12 ± 0.11	$-0.427_{(pH\,=\,5.2)}$
3.1	0.38 ± 0.06	$3.2 \cdot 10^{-4}$	3.01 ± 0.14	-0.294

$E^{0'}$ given versus SSCE at a scan rate of 20 V·s^{-1}

$$E : TcO_4^- + 2H^+ + 2e^- \xrightarrow{k_s} TcO_3^- + H_2O \qquad (4.2)$$

$$C : TcO_3^- \rightleftharpoons \text{(dehydrolysis and complexation)} \qquad (4.3)$$

$$E : Tc(V)L \rightleftharpoons Tc(VI)L + e^- + H^+ \qquad (4.4)$$

$$C : Tc(V)L \rightarrow \text{non-electroactive complexes} \qquad (4.5)$$

Several years later, Scott et al. (1989) examined the possibility of the formation of Tc-diphosphonate complex (Tc-HEDP) in aqueous solutions using a mercury electrode in a flow cell. The electroreduction of pertechnetates leads to the generation of an irreversible current wave on voltammetric curves. This wave is affected by pH and is located at ca. -0.6 V versus SSCE for pH of 4.5.

Kirchhoff et al. (1988) characterized technetium(III) complexes of trans-$[TcD_2X_2]^+$ type where D was depe or dmpe and X was Cl or Br. The experiments were carried out in aqueous solutions of KNO_3, NaCl, TEAP containing cationic, anionic and neutral surfactants. Special attention was paid on the explanation of the influence of these additives on the electrochemical signals recorded for the rhenium and technetium complexes. The authors noted that the reduction of Tc(III) complexes leads to the formation of insoluble $[Tc^{II}D_2X_2]^0$ forms, which are adsorbed at glassy carbon electrodes. CV curves recorded for these species in 0.5M KNO_3 reveal their non-Nerstian behavior. An addition of ionic surfactants (e.g., SDS or CTAB) increases the solubility of $[Tc^{II}(dmpe)_2X_2]^0$ and CV curves recorded in such solutions reveal the existence of reversible, diffusion-controlled electrochemical processes. More lipophilic $[Tc^{II}(depe)_2X_2]^0$ complexes were stronger adsorbed at the electrode surface. The standard redox potentials of $[Tc^{II}(dmpe)_2X_2]^{+/0}$ in aqueous and aqueous micellar solutions are in the range from -0.15 to -0.3 V versus Ag, AgCl (3M KCl) and for bromine complexes their values are higher than for the chloride ligands. Using chlorocoulometric measurements, the authors also analyzed diffusion of $[Re^{III}(dmpe)_2Cl_2]^+$ in 0.1M TEACl/H_2O. The respective diffusion coefficient value was found to be equal to 3.4×10^{-6} cm^2 s^{-1}. One may expect that the same value will be found also for $[Tc^{III}(dmpe)_2Cl_2]^+$. Recently, Chatterjee et al. (2013) examined luminescence properties of $[Tc(dmpe)_3]^{2+}$ complex in 0.1 KNO_3 solutions during its electrochemical reduction and subsequent oxidation, it turned out that the excited-state potentials of this complex are extremely high. Its value $(E^{\circ\prime *})$ equals 2.48 V versus SCE and is accessible using a 585 nm photon. This observation indicates that examined system has an extremely high oxidation power.

Technetium carboxylate complexes were characterized by Kennedy and Pinkerton (e.g. 1988a, b). These authors electrochemically generated Tc-formate and Tc-acetate complexes by applying a potential of -0.6 V versus SCE to a carbon electrode at pH 3–4 (Kennedy and Pinkerton 1988a). Considering the structure of other similar Tc species (see: Spitsyn et al. 1985), they concluded that such types of complexes are dioxo-bridged Tc(IV). Such types of Tc species were starting materials for chemical synthesis of Tc(III) isonitrile complexes, e.g. $[Tc(CNC(CH_3)_3)_6]^{3+}$. A discussion of their electrochemical behavior can be found in a subsequent publication of Kennedy and Pinkerton (1988b). $[Tc(CNC(CH_3)_3)_6]^{3+}$ species were reduced to Tc(I) at potentials lower than ca. -0.8 V versus SCE while its oxidation took place at 0.84 V.

De Gregori and Lobos (1989) studied electrosynthesis of technetium(V) dithiolate complexes from pertechnetates in the presence of 1,3-dimercaptopropane (DMP), 2,3-dimercaptopropanol (BAL) or 2,3-dimercaptosuccinic acid (DMSA) in alkaline (pH $= 12$) aqueous media. These species were synthesized on a mercury pool electrode by applying a potential of -1.05 V versus SCE. The authors reported that Tc(V)-dithiolato complexes were stable in the presence of oxygen and did not undergo a disproportionation reaction.

Kremer et al. (1996) studied $[(Tc$ or $Re)^V O_2(amine)_2]I$ complexes in solutions with pH varying from 2.5 to 8.5. The selected amine was a polydentate, e.g., en or tn. They concluded that TcO_2^+ cores are easier to oxidize at pH of 2.5 as compared with pH of 8.5. This is explained by the fact that at low pH values, these ions can be protonated to $TcO(OH)^+$, which undergo a very fast decomposition. Anodic branch of CV curves recorded in slightly alkaline solutions of 8.5 revealed only the signals characteristic of the iodine oxidation. The reduction of $[TcO_2(tn)_2]^+$ and $[TcO_2(en)_2]^+$ is irreversible and takes place at 0.61 and 0.54 V (vs. SHE), respectively. As one of the main conclusions of their studies, the authors stated that the stability of technetium(V) amine complexes in acid solutions turned out to be lower than for their rhenium analogs (Table 4.3).

Table 4.3 Electrochemical characteristics of selected technetium compounds (list of abbreviation and acronyms below the table)

Complex	Oxidation states and $E_{1/2}$ (or $E°$) / V	Medium	Remarks	Reference
trans-[Tc(diphos)$_2$X$_2$]$^+$	III/II X: Cl Br −0.040 0.068	0.5 M TEAP in DMF	E vs. NaSCE	(Hurst 1981)
trans-[Tc(diars)$_2$X$_2$]$^+$	III/II X: Cl Br I −0.091; −0.010; 0.105			
trans-[Tc(diars)$_2$X$_2$]	II/I X: Cl Br I −1.29; −1.15; −1.01			

trans-[Tc(dppb)$_2$Cl$_2$]$^+$
[Tc(dppe)$_2$Cl$_2$]$^+$
	III/II
	  0.092
	  −0.001
		0.5 M TEAP in DMF	E vs. Ag.AgCl (3 M NaCl); peak potential of irrev. reduction at −1.070 (or −1.080 V)	(Libson 1983)

trans-[TcD$_2$X$_2$]$^+$	III/II D: dmpe dmpe depe depe X: Cl Br Cl Br −0.231; −0.098; −0.260; −0.131 −0.190	0.5 M TEAP in DMF 0.5 M TEACl in DMF 0.5 M TEABr in DMF	E vs. Ag.AgCl (3 M NaCl); at low temp (−40 °C)	(Ichimura 1984)
trans-[TcD$_2$X$_2$]	II/I D: dmpe dmpe depe depe X: Cl Br Cl Br −1.410; −1.268; −1.439*; −1.289*	0.5 M TEAP in DMF		

(continued)

Table 4.3 (continued)

Complex	Oxidation states and $E_{1/2}$ (or $E°$) / V	Medium	Remarks	Reference
	-1.369; -1.213	0.5 M TEACl in DMF 0.5 M TEABr in DMF		
trans-[TcD_2X_2]	III/II D: dmpe dmpe depe depe X: Cl; Br; Cl Br -0.122; -0.016; -0.023; 0.115	0.5 M KNO$_3$		
[TcD_3]⁺	III/II D: dmpe depe 0.422; 0.276* 0.351; 0.175* 0.329; 0.166* 0.290	0.5 M TEAP in DMF 0.5 M TEAP in ACN 0.5 M TEAP in PC 0.5 M KNO$_3$		
trans- [Tc(PEt$_3$)$_2$(acac)$_2$en]⁺ [Tc(PEt$_2$Ph)$_2$((acac)$_2$en]⁺ [Tc(PEtPh$_2$)$_2$((acac)$_2$en]⁺ [Tc(PPh$_3$)$_2$((acac)$_2$en]⁺ [Tc(PEtPh$_2$)$_2$((bzac)$_2$en]⁺ [Tc(PEtPh$_2$)$_2$((buac)$_2$en]⁺ [Tc(PEtPh$_2$)$_2$(brac)$_2$en]⁺ [Tc(PEtPh$_2$)$_2$(sal)$_2$en]⁺	IV/III; III/II; 0.634(0.690); -1.108(-1.05) 0.674(0.710); -1.037(-0.990) 0.685(0.740); -0.946(-0.890) 0.716(0.790); -0.889(-0.840) 0.748(0.820); -0.874(-0.800) 0.619(0.650); -0.829(-0.810) 0.774(0.830); -0.751(-0.750) 0.787(0.845); -0.693(-0.670)	0.5 M TEAP in PC (0.5 M TEAP in ACN)	E vs. Ag,AgCl (3 M NaCl); $E_{1/2}$ values given in brackets are measured in ACN	(Jurisson 1984, Ichimura 1985)
trans- [Tc (dppe)$_2$(NCS)$_2$]	III/II; II/I; 0.392; -0.601	0.5 M TEAP in DMF	E vs. Ag,AgCl (3 M NaCl)	(Bandoli 1984)

(continued)

Table 4.3 (continued)

Complex	Oxidation states and $E_{1/2}$ (or E^0) / V	Medium	Remarks	Reference
[Tc H(CO)(dppe)₂] [Tc H(N₂)(dppe)₂] [Tc H(CN-tert-buthyl)(dppe)₂] [Tc H(CNC₆H₁₁)(dppe)₂]	II/I 0.59* 0.31* 0.23* 0.22*	0.2 M TBABF₄ in THF	E vs. NHE; * $E_{p/2}$ ox.; irrev. process for Tc complexes with (CO) and (N₂) ligands, quasirev. process for (CN-tert-buthyl) and(CNC₆H₁₁) ligands	(Wang 1993)
[Tc(NO)(CNCMe₃)₅]²⁺ [Tc(CNCMe₃)₆]⁺	II/I　　I/0 　　　-0.72** 0.83*	TBAP in ACN	E vs. SCE; *quasirev. oxidation; ** irrev. reduction	(Linder 1986)
[Tc(NO)(CF₃COO)₄F]²⁻	II/I -0.744V*	CF₃COOH	E vs. Fc,Fc⁺; *peak separation: 125 mV	(Balasekaran 2017)
Tc(PnAO) (ligand: PnAO-I-(2-nitroimidazole))	V/IV -1.99* 　　　　-1.52	0.1 M TBABF₄ or TBAPF₆ in DMF	E vs. Ag,Ag⁺ in ACN; * irrev. process	(Linder 1994)
TcBr₂(PMe₂Ph)₂bpy]⁺ TcCl₂(PEt₂Ph)₂bpy]⁺ TcCl₂(PMe₂Ph)₂bpy]⁺ TcCl₂(PMe₂Ph)₂phen]⁺ TcCl₂(PMe₂Ph)₂Me₂bpy]⁺ TcCl(PEtPh₂)₂terpy]⁺ TcBr(PMe₂Ph)₂terpy]⁺ TcCl(PMe₂Ph)₂terpy]⁺	IV/III;　III/II;　II/I 1.044;　-0.049; 1.080;　-0.077; 1.033;　-0.128; 1.039;　-0.130; 0.965;　-0.189; 　0.491;　-1.072 　0.467;　-1.067 　0.440;　-1.123	0.1 M TEAP in ACN	E vs. SCE;	(Wilcox 1991)

(continued)

Table 4.3 (continued)

Complex	Oxidation states and $E_{1/2}$ (or E^0) / V			Medium	Remarks	Reference
	IV/III	III/II	II/I			
[TcCl₃(PPh₃)(bpy)]	0.71;	−0.47*;	−1.43	0.1 M TBAP in ACN	E vs. SCE; * potential of irrev. peak	(Breikss 1990)
[TcCl₃(PMe₂Ph)(bpy)]	0.61;	−0.56;	−1.51*			
[TcCl₃(PMe₂Ph)(phen)]	0.61;	−0.55;	−1.49*			
[TcCl₃(PMe₂Ph)(bpm)]	0.77;	−0.40*;	−1.28*			
[TcCl₂(PMe₂Ph)₂(bpy)]⁺	1.04;	−0.13;	−1.60*			
[TcCl₂(PMe₂Ph)₂(phen)]⁺	1.05;	−0.12;	−1.59*			
[TcCl₂(PMe₂Ph)₂(bpm)]⁺	1.22;	0.05;	−1.34			
[TcCl(PMe₂Ph)₃(bpy)]⁺	1.76*;	0.30;	−0.82			
	III/II;		II/I			
trans-[Tc(SCH₃)₂(dmpe)₂]	−0.550;		−1.72	0.5 M TBAP in DMF	E vs. Ag,AgCl (3 M NaCl); Tc(III) is irrev. ox. at 0.925 V and 0.954 V for (dmpe) and (depe) respectively; for depe this process becomes rev. at low temp (−70 °C)	(Konno 1988)
-[Tc(SCH₃)₂(depe)₂]	−0.554;		−1.81			

(continued)

Table 4.3 (continued)

Complex	Oxidation states and $E_{1/2}$ (or $E°$) / V	Medium	Remarks	Reference
trans-	III/II; II/I			
-[Tc(SC$_2$H$_5$)$_2$(dmpe)$_2$]	-0.566; -1.75	0.5 M TBAP in DMF	E vs. Ag,AgCl (3 M NaCl); for trans-Tc(III) irrev. ox. at 0.87-0.92 V; this process becomes rev. for Tc-(SC$_2$H$_5$) or Tc-(S-nC$_3$H$_7$) at low temp (-70 °C); * E_{pc} or E_{pa};	(Konno 1989, 1989a, 1993)
-[Tc(S-n-C$_3$H$_7$)$_2$(dmpe)$_2$]	-0.622; -1.85			
-[Tc(SCH$_2$C$_6$H$_5$)$_2$(dmpe)$_2$]	-0.513; -1.68			
-[Tc(SCH$_2$C$_6$H$_4$-p-OCH$_3$)$_2$(dmpe)$_2$]	-0.559; -1.72			
cis-				
-[Tc(SC$_6$H$_4$-p-Cl)$_2$(dmpe)$_2$]	-0.182; -1.07*	0.5 M TEAP in DMF		
-[Tc(SC$_6$H$_5$)$_2$(dmpe)$_2$]	-0.299; -1.11*			
-[Tc(SC$_6$H$_4$-p-OCH$_3$)$_2$(dmpe)$_2$]$^+$	-0.350; -1.21*			
-[Tc(SC$_6$H$_4$-p-t-Bu)$_2$(dmpe)$_2$]$^+$	-0.382; -1.20*			
cis-	III/II; II/I			
-[Tc(SC$_6$H$_5$)$_2$(diars)$_2$]$^+$	-0.322; -1.22	0.5 M TEAP in DMF	E vs. Ag,AgCl (3 M NaCl) Tc(III) is irrev. ox. at E_{pa} = 0.94-1.08 V;	(Konno 1992)
trans-				
-[Tc(SCH$_2$C$_6$H$_5$)$_2$(diars)$_2$]$^+$	-0.362; -1.56			
-[Tc(SCH$_3$)$_2$(diars)$_2$]$^+$	-0.465; -1.70			
[Tc(tdt)(dmpe)$_2$]$^+$	IV/III; III/II; II/I 0.680; -0.600; -1.217	0.5 M TEAP in DMF	E vs. Ag,AgCl(3 M NaCl) Tc(IV)/Tc(III) couple is quasi-rev.	(Konno 1992a)
[Tc(dppbt)$_3$]	V/IV; IV/III; III/II 1.266*; 0.506*; -0.581*	0.2 M TBABF$_4$ in DCM	E vs. Fc,Fc$^+$; * potential of anodic peak; ** at -80 °C;	(Dilworth 1992)
[TcOCl(dppbt)$_2$]	-0.497**			
[M(9S3)$_2$L'$^+$]	III/II; II/I -1.08; -1.958**	0.1 M TBABF$_4$ in ACN	E vs. Fc,Fc$^+$; * quasi-rev.; ** irrev.	(White 1992, Mullen 2000)
[M(9S3)$_2$]$^{2+}$	0.87; -0.38*			

(continued)

Table 4.3 (continued)

Complex	Oxidation states and $E_{1/2}$ (or E^0) / V		Medium	Remarks	Reference
	III/II;	II/I			
[Tc(dmdc)(dppbt)₂]₊	0.312;	−0.517			
[Tc(dedc)(depe)₂]₊	0.307;	−0.544		E vs. Ag,AgCl (3 M NaCl);	(Okamoto 1993, 1993a)
[Tc(pmdc)(depe)₂]⁺ trans-	0.298;	−0.542	0.5 M TEAP in DMF	* irrev. at 25 °C;	
[Tc(SCP)₂(dmpe)₃]³⁺	−0.113;	−1.31*			
	IV/III;	III/II			
[Tc(SOS)(S-p-C₆H₄-OMe)(PMe₂Ph)]	0.175*			E vs. Fc,Fc⁺; oxidation of Tc(III) is	
[Tc(SSS)(S-p-C₆H₄-OMe)(PMe₂Ph)]	0.262*			irrev. except for Tc-(SN(Me)S)	
[Tc(SN(Me)S)(S-p-C₆H₄-OMe)(PMe₂Ph)]	0.006*			complex which is quasi-rev.	(Pietzsch 2001, 2003)
[Tc(SSS)(PPh₂S)]	0.294;	−0.468	0.1 M TBAP in DCM	* potential of anodic peak	
[Tc(SSS)(PCy₂S)]	0.200;	−0.550			
[Tc(SSS)(PMe₂S)]	0.215;	−0.565			
[Tc(SNS)(PPh₂S)]	0.144;	−0.556			
[Tc(SNS)(PMe₂S)]	0.247;	−0.568			
[Tc(SSS)(P(Me₂Ph/S)₂S)(SR)] ("3+1+1" complex 1)	0.262;	−0.335			
[Tc(SNS)(P(Me₂Ph/S)₂S)(SR)] ("3+1+1" complex 2)	0.063;	−0.520			

(continued)

Table 4.3 (continued)

Complex	Oxidation states and $E_{1/2}$ (or $E°$) / V			Medium	Remarks	Reference
	V/IV;	IV/III;				
[Tc(PPh₃)(L^1b)(morphbtu)]	1.078;	0.218		0.2 M TBAPF₆ in DCM	E vs. Fc,Fc⁺; no reduction process from -1.2 V to 0.0 V	(Huy 2009)
	V/IV;	IV/III;				
Tc[TpCF₃PC](O)	1.28;	-0.79		0.1 M TBAPF₆ in ACN	E vs. Ag,AgCl; all given redox potentials show the activity of the macrocycles	(Einrem 2016)
Tc[TpC](O)	1.18;	-0.91				
Tc[TpCH₃PC](O)	1.16;	-0.90				
Tc[TpOCH₃PC](O)	1.15;	-0.89				
trans-	IV/III	III/II				
[Tc(tmpp)₂(tmf₂en)]⁺	0.817*;	-0.556		0.1 M TEAP in PC	E vs. Ag,AgCl; *irrev.	(Baumeister 2018)
[Tc(PEt₃)₂(tmf₂en)]⁺	0.826*;	-0.567				
	IV/III	III/II	II/I			
TcCl₃(PPh₃)(CH₃CN)	1.00;	-0.49;	-2.00*;	TBAH in DMAc or DCM	E vs. NHE; * potential of irrev. cathodic peak, ** potential of irrev. cathodic peak with multielectron process	(Barrera 1996)
TcCl₃(PPh₃)(tmeda)	0.71;	-0.79*				
TcCl₃(PPh₃)(py)₃	0.79;	-0.79;	-1.72*			
TcCl₃(terpy)(DMAc)	0.67;	-0.34;	-1.42**			
TcCl₃(t-butyl₃tpy₃)	0.60;	-0.43				
[TcCl₂(Me₂Bpy)₂]OTf	1.13;	-0.16;	-1.11			
[TcCl₂(PPh₃(py)₃]OTf	0.86;	-0.53				
[TcCl₂(py)₄]OTf	1.34;	-0.38;	-1.33			
TcCl₂(py)₄		0.09;	-0.84			
[TcCl₂(py)₃]BPh₄		0.70;	-0.43			
[TcCl₂(terpy)(py)₂]OTf		0.41;	-0.82			

(continued)

Table 4.3 (continued)

Complex	Oxidation states and $E_{1/2}$ (or E^0) / V				Medium	Remarks	Reference
	IV-IV/IV-III	IV-III/III-III	III-III/III-II	III-II/II-II			
{[(terpy)(Me₂bpy)Tc]₂(μ-O)}⁴⁺/²⁺			−0.14	−0.39	TBAH in DMAc	E vs. NHE, additional one electron quasi-rev. ox. at 1.24 V and one electron irrever red. at −1.08 V	(Barrera 1996a)
(μ-O)[Cl(bpy)₂Tc]₂²⁺		0.810;	−0.584		0.1 M LiCl/H₂O		
	1.334;	0.666;	−0.773*		0.1 M TEAP in CAN		
(μ-O)[Br(bpy)₂Tc]₂²⁺		0.846;	−0.570*		0.1 M LiCl/H₂O		
	1.362;	0.703;	−0.712*		0.1 M TEAP in ACN		
(μ-O)[Cl(phen)₂Tc]₂²⁺		0.814;	−0.505		0.1 M LiCl/H₂O	E vs. Fc,Fc⁺ * potential of irrev. cathodic peak	(Lu 1993)
	1.315;	0.651;	−0.908*		0.1 M TEAP in CAN		
(μ-O)[(OH)(phen)₂Tc]₂²⁺	1.434;	0.814;	−0.676		0.1 M HNO₃		
	1.327;	0.658;	−0.727*		0.1 M TEAP in CAN		
		IV-III/III-III	III-III/III-II				
[Cl(py)₄Tc-O-Tc(py)Cl₄]		0.864;	−0.664				
[Br(py)₄Tc-O-Tc(py)Br₄]		0.737;	−0.515				
[Cl(pic)₄Tc-O-Tc(pic)Cl₄]		0.74;	−0.74			E vs. Ag,AgCl; after scaning over selected potentials were observed growing additional rev. and/or irrev. peaks	(Clarke 1988)
[Br(pic)₄Tc-O-Tc(pic)Br₄]		0.797;	−0.551		0.1 M TEAP in DMF		
[Cl(lut)₄Tc-O-Tc(lut)Cl₄]		0.840;	−0.786				
[Cl(pyr)₃ClTc-O-Tc(py)Cl₃(pyr)]		0.602;	−0.539				
[Br(py)₃BrTc-O-Tc(py)Br₃(pyr)]		0.679;	−0.298				
[Cl(pic)₃ClTc-O-Tc(pic)Cl₃(pic)]		0.56;	−0.54				
[Br(pic)₃BrTc-O-Tc(pic)Br₃(pic)]		0.612;	−0.354				

(continued)

Table 4.3 (continued)

Complex	Oxidation states and $E_{1/2}$ (or E^0) / V		Medium	Remarks	Reference
[(TCTA)Tc(μ-O)$_2$Tc(TCTA)]$^{2-}$	IV-IV/IV-III; 0.167		HOAc/NaOAc (pH 3-6); KH$_2$PO$_4$/K$_2$HPO$_4$ (pH 5-8); Tris-HCl/Tris (pH 7.5-9.5).	E vs. SCE; except reported also 2 other redox couples	(Linder 1989)
[Tc$_2$(CH$_3$CN)$_{10}$]$^{4+}$	III/II; −0.82;	II-II/II-I −1.96[*]	0.1 M TBABF$_4$ in ACN	E vs. Fc,Fc$^+$; [*] potential of cathodic peak of irrev. process; for [TcCl$_2$(CH$_3$CN)$_4$]$^+$ quasi-rev. proces at E_{pa} = −1.24 V	(Bryan 1995)
[TcCl$_2$(CH$_3$CN)$_4$]$^+$	−0.62;	−1.96[*]			
[Tc$_2$Cl$_4$(PMe$_2$Ph)$_4$]$^+$	III-III/III-II 0.92;	III-II/II-II −0.26	0.1 M TBABF$_4$ in ACN	E vs. Fc,Fc$^+$;	(Cotton 1996, 1996a)
[Tc$_2$Cl$_5$(PMe$_2$Ph)$_3$]	0.53;	−0.74			
Tc$_2$(DPhF)$_3$Cl$_2$	−0.2;	−1.5	0.1 M TBAPF$_6$ in DCM		
Tc$_2$(DPhF)$_4$Cl	−0.46;	−1.73			
[Tc$_2$(μ,η^1,η^2-CH$_3$CN)(CH$_3$CN)$_{10}$]$^{3+}$	II-II/II-I −0.25;	II-I/I-I; −1.12	0.1 M TBABF$_4$ in ACN	E vs. Fc,Fc$^+$;	(Cotton 1997)
Tc$_2$Cl$_4$(PMe$_3$)$_4$	III-III/III-II 0.894;	III-II/II-II −0.335;	0.2 M TBABF$_4$ in DCM	E vs. Fc,Fc$^+$;	(Poineau 2010)
Tc$_2$Br$_4$(PMe$_3$)$_4$	0.984;	−0.201;			

(continued)

Table 4.3 (continued)

Complex	Oxidation states and $E_{1/2}$ (or E^{o}) / V			Medium	Remarks	Reference
	V/IV					
TcOCl($L_B{}^1$)$_2$	−0.91 (at −35°C)					(Refosco 1988, Tisato 1989, 1990)
TcOCl($L_B{}^2$)$_2$	−0.87					
TcOCl($L_B{}^3$)$_2$	−0.54					
TcO(L_7)($L_B{}^1$)	−0.92			0.1 or 0.5 M TEAP in ACN or DMF	E vs. Fc,Fc$^+$	
TcO(L_7)($L_B{}^2$)	−0.88					
TcO(L_7)($L_B{}^3$)	−0.73					
TcOCl$_2$(L')	−0.69					
μ-O[TcO(L$_O$)]$_2$	−1.03					
	IV/III					
[TcL^3]$^+$	0.53			0.2 M TBAPF$_6$ in DCM	E vs. Ag/Ag$^+$; irrev.	(Hunter 1989)
	IV/III;	III/II;	II/I			
[Tc(L^1)$_3$]	0.600;	−0.997;		0.2 M TEAP in DCM	E vs. Fc,Fc$^+$; not rev.	(Refosco 1993)
[Tc(L^2)$_3$]	0.638;	−0.925;	−1.778*	0.1 M TBAP in ACN		
[Tc(L^3)$_3$]	0.756;	−0.731;	−1.881	0.1 M TBAP in ACN		
	IV/III	III/II				
[Tc(S$_2$CPh)(S$_3$CPh)$_2$]	0.894;	−0.556		0.1 M TBAP in DCM	E vs. Ag/Ag$^+$ quasirev.; irrev. ox. at 1.362 V	(Mévellec 2002)
[Tc^V(N)(FcCS$_2$)$_2$]	0.320 Fe(III)/Fe(II);				E vs. Fc,Fc$^+$; $E^{o'}$ for free ligand FcCS$_2$ equals 0.337 V; irrev. cathodic wave below −1.65 V;	(Bolzati 1997)
[Tc^V(N)(FcCS$_2$)(FcCS$_2$•)]	0.344 Fe(III)/Fe(II);			0.1 M TBAP in DCM		
	VI/VI;	VI/V;	V/IV			
Tc(DBCat)$_3$	0.45*;	−0.42;	−1.09	TBAPF$_6$ in DCM	E vs. Fc,Fc$^+$	(deLearie 1989)

(continued)

Table 4.3 (continued)

Complex	Oxidation states and $E_{1/2}$ (or E^0) / V	Medium	Remarks	Reference
Tc(DBCat)₃-(DBAP)	0.27*; −0.59; −1.53		*rev. redox system characteristic for ligands	
trans- [TcD_2X_2]⁺	III/II D: dmpe; dmpe; depe X: Cl; Br; Cl −0.323; −0.217; −0.224 −0.396; −0.300; −0.404 −0.268; −0.152; −0.248 −0.308; −0.188; −0.216	0.5 M KNO₃ 0.1 M SDS/0.1 M TEAP 32 mM CTAB/0.1 M NaCl 1.2% TritonX-100/0.1 M NaCl	E vs. Ag,AgCl (3M KCl); for KNO₃ irrev. waves; for [Tc(depe)₂Cl₂]⁺/⁰ irrev. waves	(Kirchoff 1988)
[TcO₂(amine)₂]⁺	V/IV amine: en tn 0.54* 0.61* −0.82** −0.73** −0.82** −0.73**	pH = 2.5 pH = 2.5 pH = 8.5	E vs. SHE; ionic strength 0.5 M (adjusted with KCl); irrev. waves *potential of anodic peak; **potential of catodic peak	(Kremer 1996)

(continued)

(acac)$_2$en: N,N'-ethylenebis(acetylacetone iminate); ACN: acetonitrile; (brac)$_2$en: N,N'-ethylenebis(3-bromoacetoacetone iminate); (buac)$_2$en: N,N'-ethylenebis(tert-butyl acetoacetone iminate); (bzac)$_2$en: N,N'-ethylenebis(benzoylacetone iminate); bpm: 2,2'-bipyrimidine; bpy: 2,2'-bipyridine; CTAB: hexadecyltrimethylammonium bromide; Cy: cysteine; DBAP: di-tert-butylamidophenolate; DBCat: bis(3,5-di-tert-butylcatecholate); DCM: dichloromethane; dedc: diethyldithiocarbamate; depe: 1,2-bis(diethylphosphino)ethane; diars: o-phenylenebis(dimethylarsine); diphos: 1,2-bis(diphenylphosphino)ethane; DMAc: N,N-dimethylacetamide; dmdc: dimethyldithiocarbamate; DMF: dimethylformamide; dmpe: 1,2-bis(dimethylphosphino)ethane; (H)DPhF: diphenylformamidine; dppb: bis(1,2-diphenylphosphino)benzene; dppbt(h): 2-(diphenylphosphino)benzenethiol; 2-(dppe: 1,2-bis(diphenylphosphino)ethane; en: etylenediamine; Et: ethyl; FcCS$_2$: [Fe^I(C$_5$H$_5$)(C$_5$H$_4$CS$_2$)]$^-$; FcCS$_2$•: [Fe^{II}(C$_5$H$_5$)(C$_5$H$_4$CS$_2$)]; morphbtu: N,N-dialkyl-N'-benzoylthioureas; L^1: 2-(diphenylphosphino)- benzoic; (H2L^{1b}): N'-(2-Hydroxyphenyl)benzamidine; L' = 1-(8'-quinolyliminomethyl)-2-naphtholate, L': SCH$_2$CH$_2$SCH$_2$CH$_2$S; L^2: 3-(diphenylphosphino)propionic; L^3: (diphenylphosphino)acetic; L$_B$^{1}: monoanion of N-phenylsalicylideneamine; L$_B$^{2}: monoanion of N quinolin-8-ol; L$_B$^{3}: monoanion of 5-nitroquinolin-8-ol; L$_T$: dianion of N-(2-oxidophenyl)salicylideneimine; L^3': tris[2-(2'-hydroxybenzylideneethyl)]amine; lut: 3,5-lutidine; L$_q$: N,N'-2-hydroxypropane-1,3-bis(salicylideneimine); (Me$_2$bpy): 4,4'-dimethyl-2,2'-bipyridine; PC: propylene carbonate; Ph: phenyl; phen: 1,10-phenanthroline; pic: 4-picoline; pmdc: methylenedithiocarbamate; (PnAO): derivative of propyleneamine oxime; PR$_2$S: phosphinothiolate with R = aryl, alkyl; py: pyridine; (9S3): 1,4,7-trithiacyclononane; (sal)$_2$en: N,N'-ethylenebis(salicylideneaminate); (SCP): zwitter ionic ligand −SCH$_2$P$^+$(CH$_3$)(CH$_2$)$_2$P(S)(CH$_3$)$_2$; (S$_2$CPh)(S$_3$CPh)$_2$: bis(perthiobenzoato)(dithiobenzoato); SDS: sodium dodecyl sulfate; SES: tridentate dithiolato fragment of type −S(CH$_2$)2E(CH$_2$)S− where E = O (1), S(2), NMe(3); (SR): where R can be selected alkyl or areneo- thiolato; TBAH: tetrabutylammonium hydroxide; (t-butyl)$_3$terpy): 4,4',4''-tris(tert-butyl)terpyridine; TCTA: 1,4,7-triazacyclononane-A,A',A''-triacetate; (TEA)P/Cl/Br: tetraethylammonium perchlorate/chloride/bromide; tdt: toluene-3,4-dithiolate; terpy: 2,2':6',2''-terpyridine; tmeda: etramethylethylenediamine; (tmf$_2$enH$_2$): 4,4'-[(1E,1'E)ethane-1,2-diylbis(azanylylidene)]bis(methanylylidene)]bis(2,2,5,5-tetramethyl-2,5-dihydrofuran-3-ol); tmpp: tris(3-methoxypropyl)phosphine; tn: 1,3-diaminepropane; TpXPC: mesotris(para-X-phenyl)corrole where X = CF$_3$, H, Me, and OMe

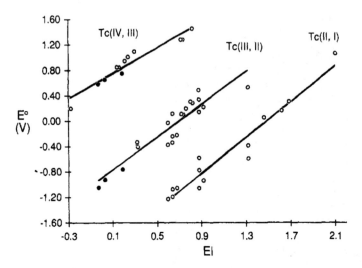

Fig. 4.5 Plot of $E°$ values for Tc^{IV}/Tc^{III}, Tc^{III}/Tc^{II} and Tc^{II}/Tc^{I} couples versus $\sum_{i=1}^{n} a_i E_L(i)$ (reprinted with permission from Lu et al. (1990) Copyright 1990 American Chemical Society)

The above described couples do not exhaust the list of the technetium complexes that can or could be used in nuclear medicine. Tc radiochemistry is a fast-growing area of modern chemistry. New Tc systems are still synthesized and characterized. Noteworthy is the fact that the half-wave potential, which is an important factor describing the stability of the complex, can be predicted for selected Tc-redox systems with a good accuracy on the basis of the properties of the ligands and the technetium core.

An in-depth analysis of the influence of the ligand on the electrochemical properties of d-elements (especially, Ru, Re and Tc) complexes was presented by Lever (1990, 1991) or Lu et al. (1990). Lever assumed that the expected standard redox potentials of many d-elements redox complexes can be calculated on a basis of a linear relationship between the ligand properties (a_i, E_L) and constants (S_M, I_M) characteristic of discussed metal redox couples (Fig. 4.5) Thus, the $E°$ values can be estimated using Eq. (4.6) where n is the coordination number and a_i is the number of donor groups in a single ligand that binds to a central atom (e.g. Tc) in a coordination complex (for example a_i for bipiridyl equals 2). S_M and I_M are, respectively, the slope and intercept of the plot for the selected technetium couples.

$$E^0 = S_M \left[\sum_{i=1}^{n} a_i E_L(i) \right] + I_M \qquad (4.6)$$

The parameters used in Eq. 4.6 for Tc(II)/Tc(I), Tc(III)/Tc(II) and Tc(IV)/Tc(III) octahedral complexes are listed in Table 4.4. In general, Lever's ligand electrochemical series predicts quite good $E^{0'}$ for various couples of Tc-complex although sometimes the calculations are incorrect as was the case with Tc complexes with pyridine ligands (see: Barriera et al. 1996).

Table 4.4 Electrochemical parameters for oxidation–reduction couples of octahedral technetium complexes in organic solvents (Lu et al. 1990)

Tc couple	Slope S_M	Std dev	Int I_M	Std dev	Corr coeff	$E^0\ M^{n+1/n+}_{(aq)}/V$
II/I	1.39	0.12	−2.07	0.10	0.95	−1.74
III/II	1.29	0.08	−0.91	0.11	0.88	−0.60
IV/III	1.00	0.10	0.64	0.09	0.92	0.89

An important conclusion arising from an analysis of the electrochemical properties of many technetium complexes is the statement indicating the impossibility of the existence of stable Tc(I) as $Tc(H_2O)_6^+$, or Tc(II) as $Tc(H_2O)_6^{2+}$ complexes. The most probable Tc(I) is not stable in any priotic medium (see: Table 4.4). One may expect that $Tc(H_2O)_6^{2+}$ should be extremely sensitive to the presence of oxygen while the standard redox potentials for Tc(III)/Tc(II) or Tc(IV)/Tc(III) couples presented in Table 4.4 indicate that $Tc(H_2O)_6^{3+}$ should be relatively stable in aqueous solutions.

Lever (1990) showed that for some isomers, e.g., cis/trans, mer/fac, of the complexes the correction leading to obtain accurate values of E^0 is necessary, especially in case of carbonyl ligands, which π^* orbitals interact with HOMO orbital of the metal complex. An earlier work of Bursten (1982) was focused on this aspect of the spectroscopic and electrochemical properties of $ML_nL'_{6-n}$ type complexes of metals with d^6 configuration (e.g. Mn(I)).

The values of E_L parameters for selected ligands are collected in Table 4.5.

Table 4.5 E_L parameters values versus NHE (Lever 1990)

Ligand	E_L/V	Ligand	E_L/V
Inorganic		Organic	
Ammonia	0.07	1,10-phenantroline (phen)	0.26
Azide(1−)	−0.30	2,9-dimethyl-1,10-phenantroline (2,9-Me₂phen)	0.20
Bromide(1−)	−0.22	4,7-dimethyl-1,10-phenantroline (4,7-Me₂phen)	0.23
Carbon monoxide	0.99	1,10-phenantroline-5,6-dione (phendione)	0.28
Chloride(1−)	−0.24	1,1,1,5,5,5-hexafluoro-2,4-pentanedionato(1−) (hfac)	0.17
Cyanate(1−)	−0.25	1,1,1-trifluoro-2,4-pentanedionato(1−) (tfac)	0.03
Cyanide(1−)	0.02	1,1-bis(diphenylarsino)methane (PDA)	0.35
Dinitrogen	0.68	1,1-bis(diphenylphosphino)methane (dppm)	0.43
Fluoride(1−)	−0.42	1,2-bis(dimethylphosphino)ethane (dmpe)	0.28
Hydride(1−)	−0.30	1,2-bis(dimethylphosphino)benzene (diphos)	0.31
Hydroxide(1−)	−0.59	1,2-bis(dimethylarsino)benzene	0.30
Iodide(1−)	−0.24	1,2-bis(diphenylarsino)benzene (diars)	0.34
Nitrate(1−)	−0.11	1,2-bis(diphenylphosphino)benzene (ophenPPh₂)	0.45
Nitrite(1−)	0.02	1,2-bis(diphenylarsinophosphino)ethane (Ph₂AsCCAsPh₂)	0.44
Perchlorate(1−)	0.06	1,2-bis(diphenylphosphino)acetylene (ADP)	0.46
Selenocyanate(1−)	−0.23	1,2-bis(diphenylphosphino)ethane (Ph₂PCCPPh₂) (dppe)	0.36
Thiocyanato(1−)	−0.06		
Water*	0.04		

(continued)

Table 4.5 (continued)

Ligand Organic	E_L/V	Ligand Organic	E_L/V
cis-1,2-bis(diphenylphosphino)ethene (Ph₂PC = CPPh₂)	0.49	1,2-bis(phenylthio)ethane	0.36
1,2-bis(diphenylphosphino)propane (PDP)	0.42	1,2-diamino-2-methylpropane	0.11
1,2-bis(ethylthio)ethane	0.32	1-(2-pyridyl)-3,5-dimethylpyrazole (pydipy)	0.23
1,2-bis(methylthlo)ethane	0.33	1,2,4-triazole, 2,3,5-tri-2′-pyridyl	0.29
1,3-diphenyl-1,3-propanedionato(1−) (dbmo)	−0.04	3-iodo-2,4-pentanedionato(1-) (3-I-acac)	−0.03
1,4,5,8-tetrazaphenanthrene	0.36	3-methyl-2,4-pentanedionato(1−) (3-Meacac)	−0.11
1-phenyl-1,3-butanedionato(1−) (bzac)	−0.06	3-phenyl-2,4-pentanedionato(1-) (3-Phacac)	−0.09
1-phenyl-4,4,4-trifluoro-2,4-pentanedionato(1−) (bztfo)	0.05	4-methoxyphenyl cyanide	0.60
1,2-bis(4-pyridyl)ethane (BPA)	0.26	4,4′-bipyridine (4,4-bpy)	0.27
2,2′-bipyridine	0.259	4,4′-bithiazole (btz)	0.20
2,2,6,6-tetramethyl-3,5-heptanedionato(1−) (dpmo)	−0.13	8-hydroxyquinolinato(1−)	−0.09
2,2′-bipyrazine (bpz)	0.36	8-methylthioquinoline	0.30
2,2′-bipyrazinium(1+)	0.75	Acetonitrile	0.34
2,2′-bipyridine, 4,4′-dibromo (4,4-Br₂bpy)	0.28	Acrylonitrile (ECN)	0.38
2,2′-bipyridine, 4,4′-dimethyl (4,4-Me₂bpy)	0.23	Benzohydroximato, p-methoxy (2−)	−0.54
2,2′-bipyridine, 4,4′-dlphenyl	0.23	Benzohydroximato, p-nitro (2−)	−0.50
2,2′-bipyridine, 4-methyl-4′-vinyl	0.23	Benzohydroximato(2−)	−0.52
2,2′-bipyridine, 4-nitro	0.30	Phenyl isocyanide	0.41
2,2′-bipyridine, 5,5′-dimethyl (5,5-Me₂bpy)	0.23	Phenyl isocyanide, 2,6-dichloro	0.46

(continued)

Table 4.5 (continued)

Ligand	E_L/V	Ligand	E_L/V
Organic		Organic	
2-(2-pyridyl)quinoline	0.25	Phenyl isocyanide, p-chloro	0.38
2,4-pentanedionato(1−)	−0.08	Phenyl isocyanide, p-methoxy	0.36
2-methylthioquinoline	0.30	Phenyl isocyanide, p-methyl	0.37
3,3′-biisoquinoline (i-biq)	0.24	Benzylamine (PMA)	0.14
3,4-bis(methylthio)toluene	0.38	Benzyl isocyanide	0.56
3,6,9-trithiaundecane	0.34	Binaphthpyridine* (binapy)	0.27
3-amino-1-propene	0.13	Bipyridazine	0.30
3-bromo-2,4-pentanedionato(1−) (3-Br-acac)	−0.03	Bipyrimidine	0.31
3-chloro-2,4-pentanedionato(1−) (3-Cl-acac)	−0.03	Bipyrimidine, 4,4-dimethyl	0.24
Biquinoline (biq)	0.29	Ethanethiolate(1−)	−0.56
Bis(4-pyridyl)acetylene	0.27	Ethyl nitrite	0.70
Bis(alkyl)-1,3-diazabutadiene	0.13	Ethylene	0.76
Bibenzimidazolato(1−)	−0.03	Ethylenediamine (en)	0.06
Biquinoline (biq)	0.29	Ethylnitrile	0.33
Bis(4-pyridyl)acetylene	0.27	Ethyl xanthato(1−)	−0.02
Bibenzimidazolato(2−)	−0.18	Formate(1−) (For)	−0.30
Bibenzimidazole (BiBizim)	0.17	Glycine(1−) (glyc)	−0.05
Biimidazole (BiimH₂)	0.13	Imidazole	0.12
1-butanethiolate(1−)	−0.55	Imidazole, 4-vinyl	0.14

(continued)

Table 4.5 (continued)

Ligand	E_L/V	Ligand	E_L/V
Organic		Organic	
Butylamine (BA)	0.13	Imidazole, N-methyl (MeIm)	0.08
Butyronitrile (PRC)	0.35	Isonicotinamide (isna)	0.26
Cyclam ([14]aneN4)	0.10	Isonitrosopropiophenoato(l-)	0.01
Cyclohexyl isocyanide	0.32	Isopropyl nitrite	0.68
Diethyldithiocarbamato(1-)	-0.08	Isopropylamine	0.05
Diethyl sulfide	0.35	Isopropyl isocyanide	0.36
Dimethyldithiocarbamato(1-)	-0.12	Maleonitriledithiolate(2-) (mnt)	-0.33
Dimethylglyoximate(1-)	0.01	Methyl nitrite	0.72
Dimethyl dimercaptomaleato(2-)	-0.47	Methyldiphenylphosphine (MPP)	0.37
Dimethylphenylphosphine (MMP)	0.34	Methyl isocyanide	0.37
Dimethylphosphine	0.34	Methyl phenyl sulfide	0.33
Dimethyl sulfide	0.31	Naphthyridine	0.24
Dimethyl sulfoxide* (DMSO)	0.47	Norbornadiene	0.46
Di-2-pyridyl ketone* (dpk)	0.28	n-butyl isocyanide	0.45
Di-2-pyridylaminato(1-)*	-0.16	N-methylbenzohydroximato, p-methyl (1-)	-0.22
Di-2-pyridylamine	0.18	N-methylbenzohydroximato, p-nitro (1-)	-0.18
N-methylpyrazinium (1+) (NMePyz)	0.79	Pyridine, 2-(aminomethyl)	0.13
N-methylpyridinium, 4-cyano (1+)*	0.45	Pyridine, 2-benzimidazolato (1-)	-0.03
N-(2-pyridylmethylidene) methylamine (pymi)	0.27	Pyridine, 2-benzimidazolyl	0.20

(continued)

Table 4.5 (continued)

Ligand	E_L/V	Ligand	E_L/V
Organic		Organic	
octaethylporphyrin(2−) (OEP) (metal in plane)	−0.07	Pyridine, 2-imidazolyl	0.18
o-acetylphenolate(1−)	−0.07	Pyridine, 2-isoquinolyl	0.26
o-propionylphenolate(1−)	−0.10	Pyridine, 2-phenylazo (Azpy)	0.40
Pentafluorobenzenethiolato(1−)	−0.33	Pyridine, 2-quinolyl	0.25
Methyldiphenylphosphine (MPP)	0.37	Pyridine, 2-tolylazo (MeAzpy)	0.41
Phenol, 2-benzimidazolato (1−)	−0.20	Pyridine, 2-(2′-naphthyridinyl)	0.22
Phenol, 2-benzimidazolato (2−)	−0.35	Pyridine, 3,5-dichloro	0.33
Phenyl cyanide	0.37	Pyridine, 3,5-dimethyl	0.21
Phenyl cyanide, 3-cyano	0.43	Pyridine, 3-(aminoethyl)	0.12
Phenyl cyanide, 4-chloro	0.40	Pyridine, 3-carboxamido	0.26
Phenyl cyanide, 4-cyano	0.49	Pyridine, 3-iodo	0.29
Phenyl cyanide, 4-methoxy	0.38	Pyridine, 4-acetyl	0.30
Phenyl cyanide, 4-methyl	0.37	Pyridine, 4-carbaldehyde	0.31
Polyvinylimidazole	0.11	Pyridine, 4-carboxamido	0.28
Pyridazine (pyd)	0.32	Pyridine, 4-carboxy	0.29
Pyrazine (pyz)	0.33	Pyridine, 4-chloro	0.26
Pyrazine, 2,3-bis(2-pyridyl)* (dpypyz)	0.32	Pyridine, 4-cyano	0.32
Pyrazole	0.20	Pyridine, 4-cyano (nitrile bonded)	0.38
Pyrazole(1-)	−0.24	Pyridine, 4-methyl (4-pic)	0.23

(continued)

Table 4.5 (continued)

Ligand	E_L/V	Ligand	E_L/V
Organic		Organic	
Pyridine	0.25	Pyridine, 4-phenyl	0.23
Pyridine, 2-(aminoethyl)	0.17	Pyridine, 4-styryl	0.23
Pyridine, 4-tert-butyl	0.23	1,2,4-triazole, 3,5-bis(pyridin-2-yl) (1−)	0.05
Pyridine, 4-vinyl	0.20	1,2,4-triazole, 3,5-bis(pyridin-2-yl)	0.11
Pyridine, poly(4-vinyl) (PVP)	0.23	1,2,4-triazole, 4-allyl	0.12
Pyrimidine (pyrim)	0.29	1,2,4-triazole(1−)	−0.17
Pyrimidinium(1+)	0.43	Triethylphosphine	0.34
Pyrrolidinecarbodithionato(1−)	−0.12	Trifluoroacetate(1-) (TFA)	−0.15
p-chlorothiophenolate(1−)	−0.43	Trifluorosulfonate(1−)	0.13
p-methylthiophenolate(1−)	−0.48	Trimethylphosphine	0.33
p-toluenethiolate(1−)	−0.48	Trimethyl phosphite	0.42
p-toluenesulfonate(1−)	−0.13	Triphenylarsine	0.38
Salicylaldehyde(1-)	−0.04	Triphenylphosphine	0.39
Dipyrido [3,2-c:2′,3′-e] pyridazine (Taphen)	0.37	Triphenyl phosphite	0.58
Terpyridine	0.25	Triphenylstibine	0.38
Tetrahydrothiophene	0.30	Tritolylphosphine (MeP)	0.37
Tetraphenylporphyrin(2−) (TPP) (metal in plane)	0.00	Tri-n-butylphosphine	0.29
Tetraaza macrocycle (TZ)	0.14	Tri-n-propylphosphine	0.34
		t-1,2-diaminocyclohexane	0.09

(continued)

Table 4.5 (continued)

Ligand	E_L/V	Ligand	E_L/V
Organic		Organic	
Thiophenolato(1−)	−0.53	t-1,2-bis(4-pyridyl)ethene (BPE)	0.26
Thiourea	−0.13	Tert-butyl isocyanide	0.36
1,2,4-triazole	0.18	Tert-butyl mercaptan(1-)	−0.55
1,2,4-triazole, 4-methyl	0.11	Vinylimidazole	0.13
1,2,4-triazole, 4-phenyl	0.14		

*E_L datum may somewhat variable from one complex to another

Table 4.6 E_L values of the ligand classes (Lever 1990)

Groups	E_L/V
OH⁻, most Xⁿ⁻ ions (inc. S anions), strong π-bases	$-0.63 \to 0$
Saturated amines falling into a fairly narrow range, weak π-acids unsaturated amines	$0 \to 0.1$
Unsaturated amines of stronger π-acid character, pyridines, bipiridines	$0.1 \to 0.40$
Hard thioethers, nitriles, soft phosphines	$0.30 \to 0.40$
Isonitriles, hard phosphines, arsine, stibines, softer phosphines	$0.35 \to 0.50$
Harder phosphines	$0.50 \to 0.65$
Dinitrogen, nitrides	$0.65 \to 0.75$
Positive-charged ligands, π-acid olefins	$0.70 \to 0.95$

An analysis of the expected values of respective redox couples potentials can be helpful in designing the strategy of the Tc complexes synthesis. As an example, such an analysis may also facilitate selection of the medium, e.g., solvent with a sufficiently wide potential window, used in the electrochemical experiments. The relative E_L values for the ligands other than listed in Table 4.5 can be estimated on the basis of their tendency to charge donation to the metal ion. This approach allows classifying the ligands groups presented in Table 4.6.

References

Armstrong R, Taube H (1976) Chemistry of trans-aquontrosyltetraamminetechnetium(I) and related studies. Inorg Chem 15(8):1904–1909

Balasekaran SM, Spandl J, Hagenbach A et al (2014) Fluoridonitrosyl complexes of technetium(I) and technetium(II). Synthesis, characterization, reactions, and DFT calculations. Inorg Chem 53:5117–5128

Balasekaran SM, Hagenbach A, Drees M et al (2017) [Tc^{II}(NO)(trifluoroacetate)$_4$F]$^{2-}$—synthesis and reactions. Dalton Trans 46:13544–13552

Bandoli G, Mazzi U, Ichimura A (1984) An isothiocyanato complex of technetium(II). Spectroelectrochemical and single-crystal X-ray structural studies on trans-[Tc(DPPE)$_2$(NCS)$_2$]^0, where DPPE =1,2- Bis (dipheny lphosphino) ethane. Inorg Chem 23:2898–2901

Barrera J, Bryan JC (1996a) Synthesis, electronic properties, and solid-state structure of {[(tpy)(Me$_2$bpy)Tc]$_2$(μ-O)}$^{4+/2+}$. Inorg Chem 35:1825–1830

Barrera J, Burrell AK, Bryan JC (1996) Technetium(III), technetium(II), and technetium(I) complexes with pyridine ligands. Can pyridine coordination stabilize the low oxidation states of technetium? Inorg Chem 35:335–341

Baumeister JE, Reinig KM, Barnes CL et al (2018) Technetium and rhenium Schiff base compounds for nuclear medicine: syntheses of rhenium analogues to 99mTc-furifosmin. Inorg Chem 57:12920–12933

Biagni Cingi M, Clemente DA, Magon L (1975) Technetium-phosphine complexes. Diethylphenylphosphonite complexes of technetium(III) and mixed ligand complexes of technetium(I) with carbonyls and diethylphenylphosphonite, and crystal and molecular structure of cis-dicarbonyltetrakis(diethylphenylphosphonite)technetium(I) perchlorate. Inorg Chim Acta 13:47–59

Bolzati C, Uccelli L, Duatti A (1997) Synthesis and characterization of nitrido Tc(V) and Re(V) complexes with ferrocenedithiocarboxylate{FcCS$_2$=[Fe(C$_5$H$_4$CS$_2$)(C$_5$H$_5$)]$^-$}. Inorg Chem 36:3582–3585

Bond AM, Colton R, Mcdonald E (1978) Chemical and electrochemical studies of tricarbonyl derivatives of manganese and rhenium. Inorg Chem 17(10):2842–2847

Breikss AI, Nicholson T, Jones AG et al (1990) Synthesis and characterization of technetium (III) and technetium (II) complexes with mixed phosphine-, chloride-, and nitrogen-donor ligands. X-ray crystal structure of TcCl$_3$(PPh$_3$)(bpy). Inorg Chem 29:640–645

Bryan JC, Cotton FA, Daniels LM (1995) Metal—metal multiply-bonded complexes of technetium. 2. Preparation and characterization of the fully solvated ditechnetium cation [Tc$_2$(CH$_3$CN)$_{10}$]$^{4+}$. Inorg Chem 34:1875–1883

Bursten BP (1982) Ligand additivity: applications to the electrochemistry and photoelectron spectroscopy of d6 octahedral complexes. J Am Chem Soc 104(5):1299–1304

Chatterjee S, Del Negro AS, Smith FN et al (2013) Photophysics and luminescence spectroelectrochemistry of [Tc(dmpe)$_3$]$^{+/2+}$ (dmpe = 1,2-bis(dimethylphosphino)ethane). J Phys Chem A 117:12749–12758

Clarke MJ, Kastner ME, Podbielski LA (1988) Low-symmetry, mixed-valent, μ-oxo technetium complexes with pyridine and halide ligands. J Am Chem Soc 110:1818–1827

Cotton FA, Haefner SC, Sattelberger AP (1996a) Metal-metal multiply-bonded complexes of technetium. 3. Preparation and characterization of phosphine complexes of technetium possessing a metal-metal bond order of 3.5. Inorg Chem 35:1831–1838

Cotton FA, Haefner SC, Sattelberger AP (1996b) Metal-metal multiply bonded complexes of technetium. 5. Tris and tetrakis(formamidinato) complexes of ditechnetium. Inorg Chem 35:7350–7357

Cotton FA, Haefner SC, Sattelberger AP (1997) Metal-metal multiply-bonded complexes of technetium. 6. A μ, η^1, η^2-CH$_3$CN complex prepared via reductive cleavage of the electron-rich Tc≡Tc triple bond in decakis-acetonitrile ditechnetium tetrafluoroborate. Inorg Chim Acta 266:55–63

Cyr JE, Linder KE, Nowotnik DP (1993) Electrochemistry of boron-capped ^{99}Tc-dioxime complexes. Inorg Chim Acta 206(1):97–104

De Gregori I, Lobos S (1989) Electrochemical and chemical reduction of TcO$_4^-$ in dithiols in alkaline aqueous media. Appl Radiat Lsot 40(5):385–391

deLearie LA, Haltiwanger RC, Pierpont CG (1989) Technetium complexes of 3,5-di-tert-butylcatechol. Direct synthesis of tris (3,5-di-tert-butylcatecholato)technetium(VI) and Bis(3,5-di-tert-butylcatecholato)- (di-teri-butylamidophenolato)technetium(VI) from ammonium pertechnetate. J Am Chem Soc 111:4324–4328

Dilworth R, Hutson AJ, Morton S et al (1992) The preparation and electrochemistry of technetium and rhenium complexes of 2-(diphenylphosphino)benzenethiol. The crystal and molecular structures of [Re(2-Ph$_2$PC$_6$H$_4$S)$_3$] and [Tc(2-Ph$_2$PC$_6$H$_4$S)$_3$]. Polyhedron 11(17):2151–2155

Einrem RF, Braband H, Fox T (2016) Synthesis and molecular structure of ^{99}Tc corroles. Chem Eur J 22:18747–18751

Hunter G, Kilcullen N (1989) Complexes of technetium, rhenium, and rhodium with sexidentate Schiff-base ligands. J Chem Soc Dalton Trans 2115–2119

Hurst RW, Heineman WR, Deutsch E (1981) Technetium electrochemistry. 1. Spectroelectrochemical studies of halogen, diars, and diphos complexes of technetium in nonaqueous media. Inorg Chem 20(10):3298–3303

Huy NH (2009) Complexes of rhenium and technetium with chelating thiourea ligands. PhD dissertation, Institute of Chemistry and Biochemistry, Freie Universität Berlin, Berlin

Ichimura A, Heineman WR, Vanderheyden J-L et al (1984) Technetium electrochemistry. 2. Electrochemical and spectroelectrochemical studies of the bis(tertiary phosphine) (D) complexes trans-[Tc^{III}D$_2$X$_2$]$^+$ (X = Cl, Br) and [Tc^{I}D$_3$]$^+$. Inorg Chem 23:1272–1278

Ichimura A, Heineman WR, Deutsch E (1985) Technetium electrochemistry. 3. Spectroelectro-chemical studies on the mixed-ligand technetium(III) complexes trans-[Tc(PR$_2$R')$_2$L]$^+$ where L is a tetradentate Schiff base ligand and PR$_2$R' is a monodentate tertiary phosphine ligand. Inorg Chem 24(14):2134–2139

Jones AG, Davison A (1982) The chemistry of technetium I, II, III and IV. Int J Appl Radiat Isot 33:867–874

Jurisson SS, Dancey K, McPartlin M et al (1984) Synthesis, characterization, and elec-trochemical properties of technetium complexes containing both tetradentate Schiff base and monodentate tertiary phosphine ligands: single-crystal structure of trans-(N, N'-ethylenebis(acetylacetone iminato))bis-(triphenylphosphine) technetium(III) hexafluorophos-phate. Inorg Chem 23(26):4743–4749

Kastner M, Fackler P, Podbielski L et al (1986) A dissymmetric μ-oxo technetium complex. Inorg Chim Acta 114: 11–L15

Kennedy CM, Pinkerton TC (1988a) Technetium carboxylate complexes-II. Structural and chemical studies. Appl Radiat Isotop 39(II):1167–1177

Kennedy CM, Pinkerton TC (1988b) Technetium carboxylate complexes-III. A new synthetic route to hexakis-(Isonitrile) technetium(I) salts. Appl Radiat Isotop 39(II):1179–1186

Kirchhoff JR, Heineman WR, Deutsch E (1987) Technetium electrochemistry. 4. Electrochem-ical and spectroelectrochemical studies on the bis(tertiary phosphine or arsine (D))rhenium(III) complexes trans-[ReD_2X_2]$^+$ (X = Cl, Br). Comparison with the technetium(III) analogues. Inorg Chem 26:3108–3113

Kirchhoff JR, Heineman WR, Deutsch E (1988) Technetium electrochemistry. 6. Electrochem-ical behavior of cationic rhenium and technetium complexes in aqueous and aqueous micellar solutions. Inorg Chem 27(20):3608–3614

Konno T, Heeg MJ, Deutsch E (1988) Thiolato-Technetium complexes. 1. Synthesis and charac-terization of bis (tertiary diphosphine) technetium (III) complexes containing methanethiolato ligands. Single-crystal structural analyses of trans-[Tc^{III}(SCH$_3$)$_2$(DMPE)$_2$]CF$_3$SO$_3$ and trans-[Tc^{III}(SCH$_3$)$_2$(DEPE)$_2$]PF$_6$, where DMPE =1,2-Bis (dimethylphosphino)ethane and DEPE = 1,2-Bis (diethylphosphino)ethane. Inorg Chem 27:4113–4121

Konno T, Kirchhoff JR, Heineman WR et al (1989a) Thiolato-technetium complexes. 2. Synthesis, characterization, electrochemistry, and spectroelectrochemistry of the technetium(III) complexes trans-[Tc(SR)$_2$(DMPE)$_2$]$^+$, where R is an alkyl or benzyl group and DMPE is l,2-bis(dimethylphosphino)ethane. Inorg Chem 28:1174–1179

Konno T, Heeg MJ, Deutsch E (1989b) Thiolato-technetium complexes. 3. Synthesis and X-ray structural studies on the geometrical isomers cis- and trans-bis(p-chlorobenzenethiolato)bis(l,2-bis(dimethylphosphino)ethane)technetium(II). Inorg Chem 28:1694–1700

Konno T, Heeg MJ, Stuckey JA et al (1992a) Thiolato-technetium complexes. 5. Synthesis, charac-terization, and electrochemical properties of bis(o-phenylenebis(dimethylarsine))technetium(II) and -technetium(III) complexes with thiolato ligands. Single-crystal structural analyses of trans-[Tc(SCH$_3$)$_2$(DIARS)$_2$]PF$_6$ and trans-[Tc(SC$_6$H$_5$)$_2$(DIARS)$_2$]^0. Inorg Chem 31:1173–1181

Konno T, Kirchhoff JR., Heeg MJ et al (1992b) Thiolato-technetium complexes part 6. Synthesis, characterization, electrochemical properties and crystal structure of [Tc(tdt)(dmpe),]PF$_6$, [tdt=toluene-3,4-dithiolate, dmpe=1,2-bis(dimethylphosphino)ethane]. J Chem Soc Dalton Trans 3069–3075

Konno T, Seeber R, Kirchhoff JR et al (1993) Thiolato-technetium complexes. 4. Synthesis, characterization and electrochemical properties of bis(1,2-bis(dimethylphosphino)-ethane)technetium(III) complexes with arene-thiolato ligands. Trans Met Chem 18:209–217

Kremer C, Domíguez S, Pérez-Sánchez M et al (1996) Comparative electrochemistry of tech-netium(v) and rhenium(v) dioxo complexes. J Radioanal Nucl Chem Letters 213(4):263–274

Lever ABP (1990) Electrochemical parametrization of metal complex redox potentials, using the ruthenium(III)/ruthenium(II) couple to generate a ligand electrochemical series. Inorg Chem 29(6):1271–1285

Lever ABP (1991) Electrochemical parametrization of rhenium redox couples. Inorg Chem 30:1980–1985

Libson K, Barnett BL, Deutsch E (1983) Synthesis, characterization, and electrochemical properties of tertiary diphosphine complexes of technetium: single-crystal structure of the prototype complex trans-[Tc(DPPE)$_2$Br$_2$]BF$_4$. Inorg Chem 22(12):1695–1704

Linder KE, Davison A, Dewan JC et al (1986) Nitrosyl complexes of technetium: synthesis and characterization of [Tc^I(NO)(CNCMe$_3$)$_5$](PF$_6$)$_2$ and Tc(NO)Br$_2$(CNCMe$_3$)$_3$ and the crystal structure of Tc(NO)Br$_2$(CNCMe$_3$)$_3$. Inorg Chem 25:2085–2089

Linder KE, Dewan JC, Davison A (1989) Technetium bis(μ-oxo) dimers of 1,4,7-triazacyclononane-N,N',N''-triacetate (TCTA). Synthesis and characterization of [(TCTA)Tc(μ-O)$_2$Tc(TCTA)]$^{n-}$ (n = 2, 3) and the crystal structure of Ba$_2$[(TCTA)Tc(μ-O)$_2$Tc(TCTA)](ClO$_4$)·9H$_2$O. Inorg Chem 28:3820–3825

Linder KE, Chan YW, Cyr JE et al (1994) TcO(PnAO-1-(2-nitroimidazole)) [BMS-181321], a new technetium-containing nitroimidazole complex for imaging hypoxia: synthesis, characterization, and xanthine oxidase-catalyzed reduction. J Med Chem 37:9–17

Lu J, Yamano A, Clarke MJ (1990) Synthesis and characterization of technetium(III) complexes with nitrogen heterocycles by O atom transfer from oxotechnetium(V) cores. Crystal structures of mer-[Cl$_3$(pic)$_3$Tc] and mer-[Cl$_3$(pic)(PMe$_2$Ph)$_2$Tc] (pic = 4-Picoline). electrochemical parameters for the reduction of TcII, TcIII, and TcIV. Inorg Chem 29:3483–3487

Lu J, Hiller CD, Clarke MJ (1993) Synthesis and electrochemistry of (g-Oxo)technetium complexes with bipyridine and halide ligands. Crystal structures of (μ-O)[X(bpy)$_2$Tc]$_2$X$_2$-bpy (X = Cl, Br) and (μ-O)[Cl(phen)$_2$Tc]$_2$Cl$_2$. Inorg Chem 32:1417–1423

Luo H (1995) Chelate complexes of rhenium and technetium: toward their potential applications as radiopharmaceutical. PhD dissertation, The University of British Columbia. Vancouver

Mazzi U, Clemente DA, Bandoli G (1977) Technetium-phosphonite complexes. Synthesis of hexacoordinate technetium(II) and technetium(III) complexes of diethyl phenylphosphonite and the crystal structure of dichlorotetrakis(diethyl phenylphosphonite)technetium(II). Inorg Chem 16(5):1042–1048

Mazzi U, Roncari E, Seeber R et al (1980) Voltammetric behaviour of technetium99 complexes with π-acceptor ligands in aprotic medium. III. Oxidation of technetium(I) complexes with phosphine and carbon monoxide ligands. Inorg Chim Acta 41:95–98

Mévellec F, Tisato F, Refosco F et al (2002) Synthesis and characterization of the "sulfur-rich" bis(perthiobenzoato)(dithiobenzoato)technetium(III) heterocomplex. Inorg Chem 41:598–601

Mullen GED, Blower PJ, Price DJ (2000) Trithiacyclononane as a ligand for potential technetium and rhenium radiopharmaceuticals: synthesis of [M(9S3)(SC$_2$H$_4$SC$_2$H$_4$S)][BF$_4$] (M) ^{99}Tc, Re, ^{188}Re) via C-S bond cleavage. Inorg Chem 39:4093–4098

Nicholson T, Mahmood A, Morgan G et al (1991) The synthesis and characterization of [M(C$_8$H$_5$N$_2$N=NH)$_3$](BPh$_4$), where M = Tc or Re. Tris-diazene chelate complexes of technetium(I) and rhenium(I). Inorg Chim Acta 179(1):53–57

Okamoto K-I, Chen B, Kirchhoff JR et al (1993a) Preparation, characterization and electrochemical properties of technetium(II) complexes with 1,2-bis(diethylphosphino)ethane (depe) and dithiocarbamate ligands. Single-crystal structural analysis of [Tc((CH$_3$)$_2$NCS$_2$)(DEPE$_2$)](PF$_6$). Polyhedron 12(12):1559–1568

Okamoto K-I, Kirchhoff JR Heineman WR et al (1993b) Preparation, characterization, electrochemical properties and single-crystal structural analysis of [Tc(SCP)$_2$(DMPE)$_2$](PF$_6$)$_3$, where SCP is the Zwitter-ionic ligand $^-$SCH$_2$P$^+$(CH$_3$)$_2$(CH$_2$)$_2$P(S)(CH$_3$)$_2$ and DMPE is 1,2-bis(dimethylphosphino)ethane. Polyhedron 12(7):749–757

Pasqualini R, Duatti A (1992) Synthesis and characterization of the new neutral myocardial imaging agent [^{99m}TcN(noet)$_2$](noet = N-Ethyl-N-ethoxydithiocarbamato). J Chem Soc Chem Commun 1354–1355

Patterson GS, Davison A, Jones AG et al (1986) Synthesis and characterization of tris(β-diketonato)technetium(III) and -(IV) complexes. Inorg Chim Acta 114:141–144

Pietzsch H-J, Tisato F, Refosco F et al (2001) Synthesis and characterization of novel trigonal bipyramidal technetium(III) mixed-ligand complexes with SES/S/P coordination (E = O, N(CH$_3$), S). Inorg Chem 40:59–64

Pietzsch H-J, Seifert S, Syhre R et al (2003) Synthesis, characterization, and biological evaluation of technetium(III) complexes with tridentate/bidentate S, E, S/P, S coordination (E) O, N(CH3), S): a novel approach to robust technetium chelates suitable for linking the metal to biomolecules. Bioconjugate Chem 14:136–143

Pihlar B (1979) Electrochemical behavior of technetium(VII) in acidic medium. J Electroanal Chem 102:351–365

Pinkerton TC, Heineman WR (1983) The electrochemical reduction of pertechnetate in aqueous hydroxyethylidene diphosphonate media. J Electroanal Chem 158:323–340

Poineau F, Forster PM, Todorova TK et al (2010) Structural, spectroscopic, and multiconfigurational quantum chemical investigations of the electron-rich metal-metal triple-bonded Tc$_2$X$_4$(PMe$_3$)$_4$ (X = Cl, Br) complexes. Inorg Chem 49:6646–6654

Refosco F, Mazzi U, Deutsch E et al (1988) Electrochemistry of oxo-technetium(V) complexes containing Schiff base and 8-quinolinol ligands. Inorg Chem 27:4121–4127

Refosco F, Tisato F, Bandol G (1993) Synthesis and characterization of neutral technetium(iii)-99 and -99m complexes with 0,P-bidentate phosphino- carboxylate ligands. Crystal structure of mer-[Tc(O$_2$CCH$_2$CH$_2$PPh$_2$),]·2Me$_2$SO. J Chem Soc Dalton Trans 2901–2908

Russell CD, Speiser AG (1982) Iminodiacetate complexes of technetium: an electrochemical study. Int J Appl Radiat Isot 33:903–906

Scott B, Schofield PJ, Deutsch EA et al (1989) Chromatographic characterization of electrochemically generated technetium-HEDP skeletal imaging agents. Talanta 36(1–2):285–292

Spitsyn VI, Kuzina AF, Oblova AA et al (1985) The chemistry of the cluster compounds of technetium. Russ Chem Rev 54(4):373–393

Tisato F, Refosco F, Mazzi U et al (1989) Synthesis, characterization and electrochemical studies on technetium(V) and rhenium(V) oxo-complexes with N, N'-2-hydroxypropane-1,3-bis(salicylideneimine). Inorg Chim Acta 64:127–135

Tisato F, Refosco F, Moresco A et al (1990) Synthesis and characterization of technetium(v) and rhenium(v) oxo-complexes with Schiff-base ligands containing the ONN donor-atom set. Molecular structure of trans- dichloro–oxo- [1-(8'-quinol yliminomethyl)-2-naphtholato-NN'O]technetium(v). J Chem Soc Dalton Trans 2225–2232

Wang Y, Pombeiro AJL, Kaden L et al (1993) Redox properties and ligand effects for the hydridotechnetium-dinitrogen, -carbonyl and –isocyanide complexes trans-(TcH(L)(Ph$_2$PCH$_2$H$_2$PPh$_2$)$_2$] (L = N$_2$, CO or CNR). In: Pombeiro AJL, McCleverly JA (eds) Molecular electrochemistry of inorganic. Bioinorganic and organomettallic compounds. Kluwer, pp 63–77

White DJ, Kuppers H-J, Edwards AJ et al (1992) Crown thioether chemistry. The first homoleptic thioether complex of technetium and its potential application in tumor imaging. Inorg Chem 31:5351–5352

Wilcox BE, Deutsch E (1991) Technetium electrochemistry. 7. Electrochemical and spectroelectrochemical studies on technetium(III) and -(II) complexes containing polypyridyl ligands. Inorg Chem 30(4):688–693

Chapter 5
Metallic Technetium, Corrosion, Technetium Alloys and Its Behavior in Spent Nuclear Fuel

Electrochemical studies on the technetium as a radioactive waste component focus on several aspects related to the waste immobilization and recycling, including:

(a) Tc reactions in nitric acid solutions;
(b) redox interactions of the technetium with other elements, such as actinides, as well as its catalytic effect on the decomposition of hydrazine and other reducing agents;
(c) Tc chemistry in alkaline aqueous solutions of high ionic strength used for the radioactive waste disposal;
(d) separation of the technetium from other elements in non-aqueous systems.

The fission process generates numerous isotopes (stable and radioactive) of various elements that are present in the nuclear fuel pellets as a contamination. Based on their different chemical properties, these elements are often classified into four groups (Kleykamp 1985a):

1. fission gases and other volatile fission products (e.g. Xe, Kr, I)
2. fission products forming oxide precipitates (e.g. Rb, Cs, Zr, Nb, Mo)
3. fission products dissolved in the fuel matrix (Sr, Nb, rare elements like Y, La, Ce, Pr, Pm, Sm)
4. fission products forming metallic precipitates (e.g. **Tc**, Rh, Ru, Pd, Ag, Cd, In, Sn, Mo).

Some of these elements may belong to more than one group. For example, molybdenum can form both metallic precipitates and oxide forms, while relatively volatile tellurium exists as oxides, cocreates metallic precipitates and can be dissolved in the fuel matrix. The tendency to the formation of oxides of selected elements present in the spent nuclear fuel (SNF) is presented by the Ellingham diagram (Kleykamp 1985a), which shows lines of coexistence of a metal and its oxide as a function of temperature. As an example of a chemical equilibrium of reactions of the fission products, one may recall the technetium–technetium oxide system (5.1).

© Springer Nature Switzerland AG 2021
M. Chotkowski and A. Czerwiński, *Electrochemistry of Technetium*,
Monographs in Electrochemistry, https://doi.org/10.1007/978-3-030-62863-5_5

Table 5.1 Fission product concentration (% in insoluble residue) in irradiated light water reactor uranium fuel (Adachi et al. 1990)

Element	Fuel burnup (MWd t^{-1} HM)				
	15 300	21 200	29 400	34 100	38 700
Mo	20	22	18.7	21.6	19.7
Tc	5	3	2.6	1.0	3.1
Ru	49	52	43.7	53.0	53.9
Rh	13	11	8.2	7.6	7.3
Pd	6	8	5.7	8.1	7.5
Amount of insoluble residues (mg)	1.6	4.7	4.7	7.0	7.0
Initial U (g)	3.336	3.037	2.094	2.153	2.059

$$Tc + O_{2(g)} \leftrightarrows TcO_2 \tag{5.1}$$

The equilibrium of oxygen pressure in this system is given by Eq. (5.2), where ΔG_m is the standard Gibbs Energy of formation of the TcO_2 as the fission product per mole of the oxygen at 1 atm. pressure and at a given temperature.

$$pO_2 = \exp\left(\frac{\Delta G_m}{RT}\right) \tag{5.2}$$

At 25 °C, $\Delta_f G_m^{\ominus}(TcO_2, cr)$ is equal to -401.8 ± 11.8 kJ mol^{-1} (Rard et al. 1999) and the equilibrium pO_2 is about 10^{-71} atm. The latter value increases with the temperature but even at 500 °C is of the order of about 10^{-41} atm. The equilibrium oxygen pressures for ruthenium, rhodium and palladium are even higher than for the technetium, which means that the former elements can exist in a metallic form even at temperatures as high as 500 °C.

Effective management of technetium as a component of the nuclear waste requires thorough understanding of fundamental chemical properties of this element. Therefore, numerous studies have been done to investigate chemical properties of various technetium forms, with a special attention paid to the pure metal, including its chemistry and electrochemistry in nitric acid solutions.

In 1968, Bramman et al. (1968) characterized the so-called "white inclusions," which were found to be essentially composed of a pseudoternary alloy Mo–(Ru–Tc)–(Rh–Pd) (Kleykamp 1985a, b). The content of these elements in the spent nuclear fuel is relatively high, as shown in Table 5.1. Their concentration in the nuclear fuel matrix increases with the burn-up of the fuel and this process is accompanied by the formation of nodules containing technetium with ruthenium, rhodium and palladium which are located along grain boundaries. These systems can exist as single hcp or bcc phases but also as di- or triphase mixtures of hcp, bcc and σ phases. Similar values of the lattice constants of Tc, Rh, Ru and Pd enable their alloying. Darby

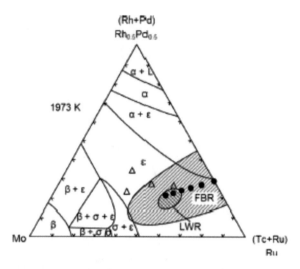

Fig. 5.1 Composition regions of the alloys (● Masahira et al. 2015, △ Yamanaka and Kurosaki 2003), together with the Mo–Ru–Rh–Pd precipitates in LWR and FBR fuels (Kleykamp 1985b), which is superimposed on the isothermal Mo–Ru–$Rh_{0.5}Pd_{0.5}$ section at 1700 °C of quaternary Mo–Ru–Rh–Pd system (Kleykamp 1985a) (reprinted with permission from Kleykamp (1985b), Masahira et al. (2015) copyright 1985, 2015 Elsevier)

et al. (1963) has shown that the technetium and ruthenium form solid solutions with hcp structures in the entire concentration range. The solubility limit of Rh and Pd in technetium at 1050 °C was found to be equal to 50 at.% and 70 at.%, respectively. A tentative phase diagram of Tc–Mo system was described by Brewer and Lamoreaux (1980).

The phase diagram shown in Fig. 5.1 reveals that Mo–(Ru–Tc)–(Rh–Pd) alloys exist in the nuclear fuel as the epsilon phase. The lattice parameters of these systems with various compositions are summarized by Rard et al. (1999).

The metallic Mo–Ru–Rh–Pd system found in SNF is dominated by the hexagonal close packing (ε) structure that occupies the bridge of the phase space, as shown in Fig. 5.1. The body centered cubic (β) space and the σ-space are limited to the Mo-rich alloys. Corrosion loss of this element makes the average composition more Pd rich. Formation of a face-centered cubic α-space is commonly observed for Pd-rich alloys obtained from high burn-up fuels.

Aihara et al. (2016) analyzed the composition of sludge remaining after dissolution of the spent nuclear fuel from "Joyo" FBR in Japan. The fuel contained from 18 to 29% of Pu and reached the burn-up from 40 to 55 GWd t^{-1} (Aihara et al. 2016). The analyzed samples were dissolved in 3.3 ÷ 11 M HNO_3, depending on the experimental procedure. The yield of the sludge was equal to 0.5 ÷ 1% of the total SNF mass. It has been found that the pseudoternary Mo–(Ru–Tc)–(Rh–Pd) alloy is mainly composed of $Mo_4Ru_4RhPdTc$. Additionally, the amount of technetium that was not dissolved in the acid and which remains in the sludge depends on the fuel composition and its burn-up and varies from 17 to 43%.

Fig. 5.2 Theoretical domains of corrosion, immunity and passivation of technetium, at 250 C (reprinted with permission from de Zoubov and Pourbaix (1966) copyright 1966 NACE International)

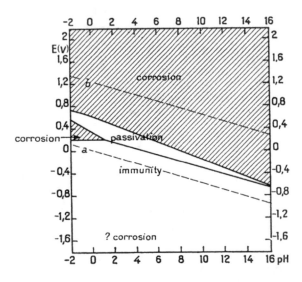

Solution obtained from dissolution of the spent nuclear fuel in HNO_3 contains the technetium mainly in the form of pertechnetates. About 10% of Tc from UOX and 30% of Tc from MOX fuels remain as a component of solid residues (Peretrukhin et al. 2008; OECD 2012). An example of the Tc mass balance in water and organic streams separated in the PUREX reprocessing of the spent nuclear fuel (burn-up $26 \div 27$ GWd t^{-1}, cooling time: $600 \div 1900$ days) was reported by Ozawa et al. (2003). About 70% of Tc is extracted to the organic phase during the first stage of extraction/scrubbing. Pu/U separation, which is the last stage of the PUREX process, generates aqueous and organic phases with practically even distribution of the technetium (6% in aqueous and 5% in organic). Although the relative technetium content in U and Pu end products is relatively low (below 0.1%) its total amount is high enough as to seriously contaminate the separated actinides.

Formation of the metallic technetium in the spent nuclear fuel and unusual properties of this element, such as relatively high resistance to the corrosion and formation of numerous alloys with noble metals drives a growing interest in studying the technetium chemistry in acidic solutions, especially those containing HNO_3.

Theoretical corrosion, immunity and passivation domains of Tc derived from its thermodynamic properties were described by de Zoubov and Pourbaix (1966) (Fig. 5.2).

The stability diagram shown in Fig. 5.2 includes only those technetium forms in which the existence was reported in 1960s. This chart is based on the Pourbaux diagram discussed in Chap. 2. Today we know that this diagram should be significantly extended by addition of other Tc forms whose existence in aqueous solutions has been recently reported, such as Tc(III), Tc(V) or Tc(IV) dimers/polymers (e.g. Rard et al. 1999). Nevertheless, the diagram quite good presents regions of the corrosion, passivity and immunity of Tc. De Zoubov and Pourbaix pointed out that the region of intensive dissolution of metallic Tc in a highly acidic environment (pH < 1) is observed at potentials higher than 0.1 V. The metallic technetium stability region is separated from the passivation region (TcO_2) by a line parallel to the hydrogen evolution line and shifted by about 0.3 V in respect to the latter. An intensive TcO_2 dissolution is observed in the entire pH range and starts above 0.76 V for pH of 0 and above -0.36 V for pH equal to 14.

The iron protection due to inhibiting action of the pertechnetates has been described in the literature (Sympson and Cartledge 1956; Cartledge 1957; de Zoubov and Pourbaix 1966; Cartledge 1971). Cartledge (1955) assumed that the addition of TcO_4^- ions at a level of 1 mM ($KTcO_4$) has a significant effect on the corrosion rate of Fe. He pointed out that the passivation action is related to the feeble reversible adsorption of TcO_4^- ions to the iron. The inhibitory properties are also attributed to TcO_2, which is formed as a result of the reduction of TcO_4^- on the surface of the metal. Cartledge also indicated that mild carbon steel can be protected against corrosion in aerated distilled water even up to 250 °C when the latter contains $5 \div 50$ ppm of Tc. His later work (Cartledge 1971) concerned on analysis of the corrosion process of metallic Tc and Tc (0.1% w/o)-Fe alloy in H_2SO_4/Na_2SO_4 solutions (pH of 0.35 or 2.7). For the Tc–Fe alloy Cartledge obtained the Tafel slope of 110 mV/decade. The rate of the oxidation of the pure metal or its oxides was estimated at a level of μA cm^{-2}. Surface of the fresh Tc–Fe electrode exhibits uniform Tc distribution with the average concentration of 10^{12} atom cm^{-2} while the anodic polarization at 4.6×10^{-2} A cm^{-2} leads to the surface enrichment with Tc up to $10^{16} \div 10^{17}$ atom cm^{-2}. Such surface enrichment results in a decrease in the hydrogen evolution overpotential and ennobles the open circuit potential.

Cartledge (1971) analyzed hydrogen evolution on a metallic technetium electrode in a 0.5 M H_2SO_4 solution and determined the exchange current density of 9.5×10^{-5} A cm^{-2}. Trasatti (1972b) suggested a slightly higher value of this parameter (1.6×10^{-4} A cm^{-2}) and this range of 10^{-4} A cm^{-2} was recalled in a more recent work of Jaksic (Jaksic 2000). Other than activity toward the hydrogen evolution, fundamental electrochemical properties of the metallic technetium are discussed below. The experimental value of the work function (ϕ) of polycrystalline metallic technetium is equal to 4.58 eV. Theoretical calculations for hcp (001) and fcc (111) surfaces of Tc yield the ϕ values of 5.36 eV and 5.42 eV, respectively (Skriver 1992). Other Tc surfaces for which the work function has been reported include (0001) Tc (4.66 eV), Tc (10-10) prismatic (4.48 eV) and Tc (10-10) basal (4.19 eV) (Taylor 2011).

Taylor (2011) reported the results of computational studies on H and O adsorption at the surface of Tc and a Tc–Fe alloy. The density functional theory was used to calculate its adsorption energies. For the oxygen, this parameter was found to depend on the technetium surface geometry and equals -3.67 eV for Tc (basal), -3.47 eV for Tc (prismatic) -3.48 eV for Tc (pyramidal). The author assumed that these values are higher than those found for the same processes taking place on a TcO_2 surface. An E–pH diagram illustrating O, OH and H adsorption regions is presented in Fig. 5.3.

Zhuz and Wang (2016) studied halides adsorption on a Tc surface (Zhuz and Wang 2016). This process leads to a change in the work function ($\Delta\phi$) of hcp Tc surfaces with the $\Delta\phi$ increasing in the order F < Cl < Br < I and varying from 0.15 eV for

Fig. 5.3 Potential pH diagram illustrating regions of Tc stability regarding the water dissociation and hydrogen adsorption reaction (Taylor 2011). The surface adsorption regions are superposed upon the bulk phases obtained from Pourbaix (reprinted with permission from Taylor (2011) Copyright 2011 Elsevier)

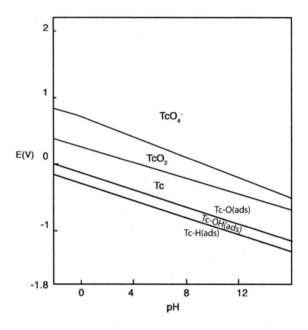

F to -0.31 eV for I. The fcc sites exhibit higher work function changes due to the halides adsorption with $\Delta\phi$ varying from 0.71 eV to -0.31 eV for the F/Tc and I/Tc systems, respectively. ΔE_{ads} of adsorption of the halides on a hcp Tc site is equal to -4.74, -3.97, -3.58 and -3.42 eV for F, Cl, Br and I, respectively.

Metallic technetium can be obtained by electroreduction of the TcO_4^- ions in acidic solutions. An overview of various types of galvanizing baths used to deposit this element can be found in a work by Voltz and Holt (1967). The authors stated that a shiny metallic cathodic deposit can be obtained from H_2SO_4 solutions with a low pH. The authors noted that during the process, the initially colorless solutions are turned into colored liquids indicating formation of ionic forms of the technetium. Tc oxide coatings were deposited from NaOH solutions (pH = 12) or from H_2SO_4 (or NH_{3aq}) with the addition of citric acid (pH = 2 ÷ 9). A decrease in the current density from 4 to 1 A dm^{-2} resulted in an increase in the cathodic current efficiency (CCE) from 11 to 27%. An elevation of the electroplating bath temperature from 25 to 70 °C results in the CCE increase from 18 to 26%. Voltz and Holt have found that the optimal composition of technetium electrodeposition bath was 1 M $(NH_4)_2SO_4$ and 18 mM NH_4TcO_4 with pH of about 1 (by addition of H_2SO_4). The electrolysis of such type of the solution with a current of 1 ÷ 2 A dm^{-2} leads to the formation of a bright deposit. The efficiency of this process was 18 ÷ 30%.

Box (1968) described the methods of Tc deposition on large surfaces from 0.7 M ammonium oxalate and sulfuric acid solutions containing pertechnetates at current densities of 100 ÷ 130 A dm^{-2}. The H_2SO_4 concentration varied depending on the cathode material, e.g., from 0.45 mol dm^{-3} for Cu to 1.90 mol dm^{-3} for Pt. The Tc electrodeposition yield was high, above 99% and the deposits of up to 18 mg cm^{-2} were obtained.

A similar to Box (1968) composition of electrodeposition bath was described by Wotteen (1977). The bath contained 0.7 M ammonium oxalate and 1.4 M sulfuric acid and its pH was equal to 1. Tc was deposited as a protective coating of stainless steel using a current density of 1.3 A cm^{-2}.

Electrodeposition from bath with high pH leads to the formation of the technetium oxide instead of pure metal (Boyd 1959).

Zakharov et al. (1991) deposited technetium on Ni electrodes from a bath with pH of 1 using currents of 7 or 12 A dm^{-2}. He determined on the basis of crystallographic studies that such obtained deposit was a mixture of crystalline and amorphous phases. He established that technetium hydride with maximum hydrogen content corresponding to $TcH_{x<0.27}$ and with lattice constants $a = b = 2.793 \pm 0.002$ Å and $c = 4.444 \pm 0.002$ Å can be formed during this process.

Mausolf et al. (2011), in turn, deposited metallic technetium from 1 M H_2SO_4 containing 2 or 10 mM TcO_4^- and applying 1.0 A×cm^{-2} and various deposition times from 100 to 2000s. They observed that the deposition of the metal is hindered by generation of Tc(IV) species characterized spectroscopically by the band near 500 nm. The efficiency was below 10% for 2000s of the Tc electroplating.

Deposition of technetium from 1 M H_2SO_4 solutions was discussed also by Engelmann et al. (2008). The electrolysis was carried out in a two-electrode system with a platinum disk acting as a cathode. The technetium was deposited from 0.25 ml solutions containing 25 ng of Tc at the voltage of 5 V (~20 mA). The deposition efficiency varied from 69 to 84% for 6 and 18.6 h of the electrolysis, respectively. The electrolysis with the current of 0.1 ÷ 0.5 A carried out for 1.5–4 h led to the formation of the Tc coatings with the yields from 20 to 48%.

Boyd et al. (1960) investigated codeposition of technetium and rhenium from slightly acidic solutions. He concluded that the electroplating can be effectively carried out at pH of 5.5 and in the presence of 2 mM F^-. The electrolysis of 15 ml of the solution at 100 mA cm^{-2} for 2 h with a copper cathode yielded the Tc deposition at a level of 89%. The deposit obtained under such experimental conditions was most likely composed with a hydrated TcO_2.

Metallic technetium can be obtained also in concentrated potassium acetate or mixed potassium and ammonium acetate solutions at potentials lower than -1.4 V or -1.15 V (vs. saturated Ag, AgCl reference electrodes) respectively (Kuznetsov 2020). This process is multistep and occurs via generation of Tc(IV) and Tc(III) forms. An earlier work of Maslennikov et al. (1998) has shown that metallic technetium can be electrodeposited with relatively high efficiency (92–95%) from aqueous formate solutions (pH = 6.0–7.5) by applying potential lower than -1.4 V versus SCE.

A summary of the described baths used for the technetium deposition from aqueous solutions is provided in Table 5.2.

A procedure to fabricate smooth coatings of transition metals (Cr, Zr, Hf, W, Mo, Mn, Tc, Re or lanthanides) with a 10-point average roughness below μm using electrodeposition from molten salt baths was patented by Inazawa et al. (2016). The

Table 5.2 Selected baths for the electroplating of technetium

Solution	Applied current	Electrode	Remarks	References
0.7 M ammonium oxalate + H_2SO_4	$100 \div 130$ A dm^{-2}	Cu	0.45 M H_2SO_4, metallic deposit	Box (1968)
		Ni	0.87 M H_2SO_4, metallic deposit	
		Al	1.14 M H_2SO_4, metallic deposit	
		Ag	1.41 M H_2SO_4, metallic deposit	
		Au	1.41 M H_2SO_4, metallic deposit	
		Stainless steel	1.41 M H_2SO_4, metallic deposit	
		Pt	1.90 M H_2SO_4, metallic deposit	
H_2SO_4 H_2SO_4 1 M H_2SO_4 2 M H_2SO_4 H_2SO_4 + NH_4HF_2 H_2SO_4 + 1 M $(NH_4)_2SO_4$ H_2SO_4 + Na_2SO_4 H_2SO_4 + K_2SO_4	$1 \div 4$ A dm^{-2}	Au, Cu, Stainless steel	pH = 2.0, deposit dark, loose, bath darkened pH = 1.0, deposit shiny, metallic, bath darkened deposit shiny, metallic, bath darkened deposit shiny, metallic, bath darkened pH = 1.0, deposit shiny, metallic, bath darkened, pH = 0.5 ÷ 1.5, deposit shiny, metallic, black when concentration of $(NH_4)_2SO_4$ is above 1 mol dm^{-3}, bath turned light pink pH = 1.0 deposit shiny, metallic, bath light pink pH = 1.0, deposit shiny, metallic, bath light pink	Voltz and Holt (1967)

(continued)

Table 5.2 (continued)

Solution	Applied current	Electrode	Remarks	References
NH_4TcO_4	7 or 12 A dm^{-2}	Ni	pH = 1, XRD study of technetium hydride, $TcH_{<0.27}$	Zakharov et al. (1991)
1 M H_2SO_4	5 V (~20 mA)	Pt	Trace amounts of Tc electrodeposited with 69÷84% efficiency	Engelmann et al. (2008)
2 M H_2SO_4	1 A cm^{-2}	Au	XAFS, XRD, SEM study of technetium deposit bath darkened, formation of Tc(IV) species	Mausolf et al. (2011)
8 M CH_3COOK or 4 M CH_3COONH_4 4 M CH_3COOK	5 mA cm^{-2}	Cu	pH = 7, XANES/EXAFS, SEM study of technetium deposit, formation of dinuclear Tc(IV,III) species	Kuznetsov et al. (2020)

use of a mixture of halides of alkaline metals or alkaline earth metals (or beryllium) enables to carry out the electrodeposition of the metals below 400 °C. An important addition to the bath is an organic polymer with a weight-average molecular weight of at least 3 000 at 0.0001 ÷ 1 mass%. Its task is to block the sharp edges of the surface of the substrate by adsorption on them. This results in smooth deposits of the transition metals. An exemplary bath used by the inventors for electrodeposition of chromium at 250 °C contained 56.1 LiBr, 18.9 KBr, 25.0 CsBr and 2.78 $CrCl_2$ (in % mol ratio) and polyethylene glycol (PEG) with the mass of 20,000 or polyethylene imide (PEI) with the mass of 100,000 or 750,000. The concentration of both polymers varied from 0.0195 to 0.0955%. The cathode potential was 50 mV lower than the threshold value. According to this patent application, the deposition process can be effectively accomplished regardless on the chemical form of the technetium and its content in the molten salt.

Unfortunately, the literature description of the electrochemical properties of the metallic technetium is very incomplete and selective. Cartledge (e.g. Cartledge 1971) was the only one who widely discussed hydrogen evolution on the technetium. He obtained the Tafel slope of approximately 40 mV decade^{-1}. According to Trasatti (1972a), this value indicates a fast proton discharge followed by a rate-determining step, which was identified as a reaction of hydrogen ion with adsorbed hydrogen atom.

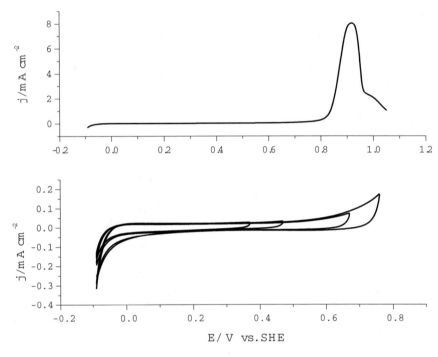

Fig. 5.4 Second cycle of the voltammetry of initially electrodeposited of Tc on Au surface ($t_{electrodep}$ = 1 h, j = 4 mA cm^{-2}, TcO$_4^-$ = 0.16 mM) in 0.5 M H$_2$SO$_4$ at a scan rate of 5 mV s^{-1} and various anodic vertex potentials

The electrochemical properties of the metallic technetium reported in the literature so far are limited almost exclusively to its dissolution under various experimental conditions. Much more data are available for rhenium, which is a Tc analog, and the results reported for the former are often used to deduce behavior of the technetium. Hydrogen evolution on rhenium was covered by several recent publications (e.g. Garcia-Garcia et al. 2014, 2016). It is reported that the process proceeds via the mechanism with the fastest reaction rates, Eqs. (5.3) and (5.4):

$$M + H^+ + e^- \leftrightarrows MH_{ads} \qquad (5.3)$$

$$MH_{ads} + H^+ + e^- \leftrightarrows M + H_2 \qquad (5.4)$$

The same reaction scheme is expected to be valid also for the technetium.

Figure 5.4 shows typical CVs recorded for the metallic technetium in acidic solutions. A strong increase in cathodic current appears below approx. −0.05 V. Current signal characteristics of deposited metallic technetium in acidic solutions are similar to those observed for the metallic rhenium (Garcia-Garcia et al. 2014). Similarly to Re and Ru, the voltammetric curves recorded for the metallic Tc do not reveal the

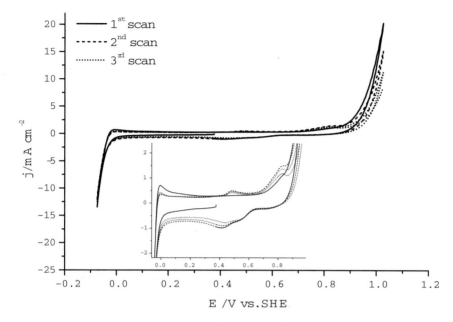

Fig. 5.5 Cyclic voltammetry of Tc on Au surface ($t_{electrodep} = 1$ h, $j = 1$ A cm^{-2}, TcO$_4^- = 0.16$ mM) in 0.5 M H$_2$SO$_4$ at a scan rate of 5 mV s^{-1} and various anodic vertex potentials

most probable well characterized double layer charging region. An increase in the anodic currents observed at potentials above c.a. 0.6 V is attributed to oxidation of the Tc. The latter process results in the generation of the technetium dioxide and, finally, the pertechnetates. This oxidation onset is ca. 0.1 V higher as compared to the metallic rhenium and this potential shift correlates well with the difference in the standard redox potentials of ReO$_4^-$/Re ($E^\ominus = 0.369$ V) and TcO$_4^-$/Tc ($E^\ominus = 0.472$ V) couples (Magee 1974). It should be stressed, however, that the rhenium oxidation may result in formation of not only the perrhenates but also of numerous oxides, e.g. Re$_2$O$_5$ or ReO$_3$ (Szabó and Bakos 2000). In contrast to the hydrated TcO$_2$ (Rard et al. 1999), which is the only one Tc oxide whose existence in aqueous solutions is undoubtedly confirmed, the structure and properties of these rhenium oxides are well resolved.

Electrodeposition with current density as high as $j = 1$ A cm^{-2} (Mausolf 2011) generates the metallic Tc deposits thicker than those obtained for the same deposition time but for lower currents (mA cm^{-2} range). It is worth mentioning that the current densities shown in Figs. 5.4 and 5.5 are expressed in respect to the real surface area of the gold.

Cyclic voltammetry curves shown in Fig. 5.5 indicate that the oxidation of the metallic Tc is fast at potentials higher than 0.8 V. It is worth noting that the first polarization cycle recorded in the anodic direction does not reveal currents due to oxidation of reduced forms of technetium (e.g. Tc(III) to Tc(IV)), which should appear at ca. 0.48 V. This signal is observed only for subsequent cycles. At a potential

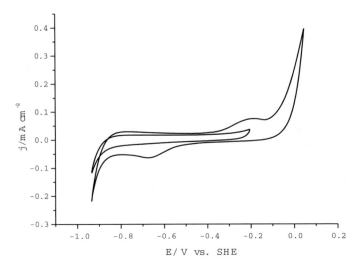

Fig. 5.6 Cyclic voltammetry of Tc on Au surface ($t_{electrode} = 1$ h, $j = 1$ A cm^{-2}, TcO$_4^-$ = 0.16 mM) in 0.1 M NaOH at a scan rate of 5 mV s^{-1} and various anodic vertex potentials; second scan

of about 0.82 V, a small peak is formed, which is probably due to oxidation of technetium(IV) polymeric forms or Tc oxides to TcO$_4^-$. The overall anodic current decreases with the number of potential scan cycles. Such a behavior indicates the effective dissolution of the technetium deposit with formation of the pertechnetates and simultaneous reduction of the active Tc surface area. The pertechnetates are electroreduced during the cathodic potential scan below 0.6 V. The evolution of this signal seen for second and third cycles seen in Fig. 5.5 resembles changes of the currents of the pertechnetates reduction recorded on the gold electrode in 0.5 mM TcO$_4^-$ + 1 M H$_2$SO$_4$ (see: Chotkowski and Czerwiński 2012).

In contrast to the acidic environment, the technetium electrochemistry in alkaline solutions has not been described in the literature in greater detail so far. Typical CVs recorded for this metal in alkaline solutions are presented in Fig. 5.6. Intensive dissolution of the technetium deposit is observed at potentials positive to −0.2 V. Similarly to the acidic solutions case, dissolution in alkaline electrolytes also leads to the formation of the pertechnates as the ultimate product of the reaction. It cannot be excluded, however, that Tc(V) and/or Tc(VI) species are also generated during this process. The cathodic branch of the CV curve reveals a broad reduction wave at potentials lower than −0.6 V, which is attributed to reduction of initially generated TcO$_4^-$ to mainly Tc(IV).

It is worth to direct the reader's attention to the current density values shown in Figs. 5.5 and 5.6. They differ significantly depending on the electrode history. We pay attention to this methodological aspect of working with the metallic technetium. The technetium is one of the metals for which the accurate and reliable electrochemical methods of "in situ" real surface area determination have not been developed so far. Thus, the currents in the figures in question are expressed in respect to the real surface

Fig. 5.7 Cyclic
voltammetric curves at a
Tc-covered surface ($\Gamma = 2 \times 10^{-8}$ mol cm^{-2}) in 1 M
H$_2$SO$_4$ (1) and in 3 M
HClO$_4$ (2). Sweep rate,
2.5 mV s^{-1}, (geometric
surface area: 13 cm^2)
(reprinted with permission
from Láng and Horányi
(2003) copyright 2003
Elsevier)

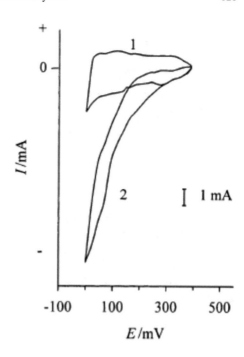

area of the gold substrate used for the Tc electrodeposition (according to Trasatti and Petri 1991).

Interesting properties of the technetium coatings have been described by Horányi et al. (e.g. Láng and Horányi 2003). Figure 5.7 shows the cyclic voltammograms of an inert electrode covered by a Tc deposit. A significant increase in the reduction currents is observed when a H$_2$SO$_4$ electrolyte is replaced with a HClO$_4$ solution. The authors attribute this effect to the catalytic electroreduction of ClO$_4^-$ ions at the Tc surface.

Ferrier et al. (2013) investigated the dissolution of metallic technetium at various potentials in solutions containing acids (1 ÷ 6 M HCl or HNO$_3$) and salts (1 M NaCl or NaNO$_3$). In general, this process is the most effective in 1 M HNO$_3$. At potential of 0.8 V (vs. Ag, AgCl, KCl sat.), its rate is equal to 1.06×10^{-4} mol cm^{-2} h^{-1}. An analysis of the Tafel slopes revealed that the technetium dissolution potential is equal to 0.596, 0.601 and 0.832 V in 1, 2 and 6 M HNO$_3$ solutions, respectively. TcO$_4^-$ ions are the only products of the corrosion in HCl and HNO$_3$ solutions. When the electrode potential is relatively low (0.7 or 0.8 V), the rate of the Tc dissolution in HCl solutions is higher than in HNO$_3$. This trend is reversed when the potential becomes as high as 1 V. These authors also noted that the Tc dissolution rate at low potentials in NaCl and NaNO$_3$ solutions with pH of 1 or 2.5 was higher than in HCl or HNO$_3$ solutions. One of the recent works of Kenneth Czerwinski's group (Poineau et al. 2016) deals with the studies on the solubility of Tc–Ru alloys in nitric acid solutions. The technetium dissolves in these solutions at potentials higher than about

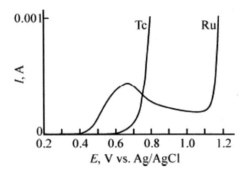

0.7 V. The nobler ruthenium undergoes effective dissolution at potentials higher than about 1.1 V (Fig. 5.8).

A gradual increase in the ruthenium content in the Tc–Ru alloy improves its resistance to dissolution. The authors reported the following relation between the transpassivation potential of the Tc–Ru alloy and the ruthenium content in it, Eq. (5.5):

$$E_{tp} = 1.16 - 0.492 \times \left(0.963^{[Ru]}\right) \tag{5.5}$$

The ^{99}Tc is a long-lived beta minus emitter and slowly decays to a stable ^{99}Ru isotope. An increase in the transpassivation potential of this system is therefore predicted and after 1 half-life of ^{99}Tc its value in 1 M HNO₃ should reach 1.09 V (vs. Ag, AgCl) as compared with 0.67 V observed for the metallic technetium.

Tc–Ru alloys with ruthenium content of 19, 50 and 70% at were investigated also by Maslennikov (2012). Formation of passive films containing Tc(IV)–Ru(III,IV) hydroxides was observed at potentials lower than 0.65 V in HNO₃ solutions with concentration below 2 mol dm⁻³ (vs. Ag, AgCl). The rate of dissolution of the metallic Tc under open-circuit conditions varied from 2.9 μg cm⁻² h⁻¹ in 0.5 M HNO₃ to 3.3 mg cm⁻² h⁻¹ in 6 M HNO₃. The effect of the Ru content on the corrosion potential of Tc–Ru alloys can be traced by analyzing Table 5.3. Maslennikov observed a decrease of the exchange current values with the increase in the Ru concentration in the alloy. The highest decrease was observed in 6 M HNO₃. The values of the transpassivation potential also increased with the Ru concentration in the alloy and this effect was especially evident in 6 M HNO₃. The author concluded that the addition of Ru to the Tc–Ru alloy increases significantly the stability of the latter in the nitric acid solutions.

The dissolution of the metallic technetium in 0.5 ÷ 6 M HNO₃ was further investigated by Rotmanov et al. (2015). These authors reported faradaic efficiencies greater than 100% for current densities from 0.86 to 319 mA cm⁻², which indicates that a chemical dissolution also contributes to the overall process. Microscopic studies revealed that corrosion degradation of the metallic Tc has an intercrystalline character. Spectroscopic measurements showed that this process is accompanied by the

Table 5.3 Corrosion and dissolution characteristics of Tc metal and its alloys with Ru in 0.5 ÷ 4 M HNO$_3$ (Maslennikov 2012). The errors in the Tc–Ru composition in the origin table have been corrected based on the text of the article

Electrode	HNO$_3$/ mol dm^{-3}	E_{corr}/ V versus Ag, AgCl	j_0/ μA cm^{-2}	E_{tr}/ V versus Ag, AgCl
Tc metal	0.5	0.5 ± 0.06	0.4 ± 0.2	0.69 ± 0.01
	1.0	0.56 ± 0.04	0.7 ± 0.3	0.66 ± 0.02
	2.0	0.63 ± 0.02	2 ± 3	0.67 ± 0.02
	4.0	0.73 ± 0.01	70 ± 10	0.73 ± 0.01
	6.0	0.810 ± 0.005	600 ± 100	–
Tc-19%Ru alloy	0.5	0.41 ± 0.04	0.6 ± 0.4	0.97 ± 0.01
	1.0	0.56 ± 0.04	1.3 ± 1	0.92 ± 0.02
	2.0	0.62 ± 0.01	1.3 ± 0.4	0.98 ± 0.01
	4.0	0.80 ± 0.01	0.7 ± 0.4	1 ± 0.01
	6.0	0.85 ± 0.01	2 ± 1	0.98 ± 0.01
Tc-50%Ru alloy	0.5	0.48 ± 0.06	1 ± 0.3	1.06 ± 0.03
	1.0	0.53 ± 0.05	0.4 ± 0.3	1.09 ± 0.01
	2.0	0.56 ± 0.05	0.4 ± 0.3	1.09 ± 0.01
	4.0	0.82 ± 0.02	0.5 ± 0.3	1.12 ± 0.06
	6.0	0.86 ± 0.01	0.6 ± 0.3	1.12 ± 0.01
Tc-70%Ru alloy	0.5	0.47 ± 0.03	0.4 ± 0.1	1.13 ± 0.02
	1.0	0.53 ± 0.01	0.4 ± 0.06	1.13 ± 0.01
	2.0	0.63 ± 0.04	0.6 ± 0.3	1.17 ± 0.01
	4.0	0.81 ± 0.05	0.7 ± 0.2	1.18 ± 0.01
	6.0	0.87 ± 0.01	0.6 ± 0.2	1.21 ± 0.01

formation of intermediate species characterized by the wave at λ ~ 480 nm, which can be most likely attributed to Tc(V).

Poineau et al. (2014) studied electrochemical behavior of Tc–Ni systems containing 1 and 10% of technetium. They concluded that an addition of 10 wt% of technetium to Ni does not have significant influence on the open circuit potential of the system, which has a value typical for a crude Ni. Kolman et al. (2012), in turn, investigated the corrosion of Tc–Fe alloys in a broad range of technetium content (0, 10, 50, 70, 100%). The examined alloys were made by arc melting followed by annealing for 4 h at 1600 °C in Ar atmosphere in a resistance furnace. The authors did not observe passivation of the investigated alloys in 0.1 mM H$_2$SO$_4$ although the results for the 10 wt% Tc sample cannot be unequivocally interpreted due to significant ohmic loss in the solution.

The studies on the pertechnetates electroreduction in HNO$_3$ solutions were conducted also by Zilberman et al. (2007). They analyzed solutions of 0.5 M HNO$_3$

containing 0.1 or 1 g dm^{-3} TcO$_4^-$ using Au, GC and Pt electrodes under hydrodynamic conditions (RDE). The authors stated that the currents due to the pertechnetate
reactions were indistinguishable from the TcO$_4^-$ free background.

The experiments carried out by Bebko (2011) showed that even small addition of
HNO$_3$ to a solution containing the pertechnetates significantly affects the currents
due to TcO$_4^-$ reactions. A large, distorted reduction wave was observed in 4 M
H$_2$SO$_4$ solutions containing 50 mM HNO$_3$ and 0.32 mM TcO$_4^-$ at potentials below
0.2 V (vs. Hg, Hg$_2$SO$_4$, 0.5 M H$_2$SO$_4$). This signal contains charge associated with
the electrochemical reduction of TcO$_4^-$ but the main contribution comes from electrochemical degradation of HNO$_3$. Additional spectroelectrochemical measurements
confirmed the formation of HNO$_2$ and NO$_x$.

Before starting more precise characterization of the catalytic properties of Tc in
HNO$_3$ solutions, it is necessary to discuss possible redox couples containing NO$_x$,
which may exist in the nitric acid solutions. HNO$_3$–HNO$_2$–NO$_x$ systems exhibit
relatively high values of the standard redox potentials (Compton and Sanders 1996)
(5.6–5.8):

$$NO_3^- + 4H^+ + 3e^- \leftrightarrows HNO_2 + H_2O \quad E^\ominus = 0.94\,V \qquad (5.6)$$

$$NO_3^- + 4H^+ + 3e^- \leftrightarrows NO + 2H_2O \quad E^\ominus = 0.96\,V \qquad (5.7)$$

$$2NO_3^- + 4H^+ + 2e^- \leftrightarrows N_2O_4 + 2H_2O \quad E^\ominus = 0.81\,V \qquad (5.8)$$

The above standard redox potential values are higher than those for the reduced
technetium species (Rard et al. 1999) which means that the latter may be involved
in redox reactions with the nitrates. A further complication in the analysis of the
HNO$_3$–Tc systems comes from the decomposition of HNO$_3$ accelerated by the
accumulation of HNO$_2$ (5.9):

$$NO_3^- + HNO_2 + H^+ \leftrightarrows 2NO_2 + H_2O \qquad (5.9)$$

Unstable HNO$_2$ shows strong oxidizing properties (5.10):

$$HNO_2 + H^+ + e^- \leftrightarrows NO + H_2O \quad E^\ominus = 0.99\,V \qquad (5.10)$$

and is able to oxidize the reduced Tc.

Chotkowski and Czerwiński (2016) investigated the stability of technetium
species with intermediate oxidation states of in the presence of nitric acid. The solutions containing initially electrochemically generated polymeric Tc(III/IV) species
were acidified by addition of HNO$_3$. Oxidation of these Tc forms was accompanied
by generation of a new band at 465 nm (Fig. 5.9). The maximum of this wave is close
to the value of 480 nm assigned to the Tc(V) ions by Rotmanov (2015). This wave is
broad and poorly shaped, which suggests that apart from Tc(V) also other Tc species

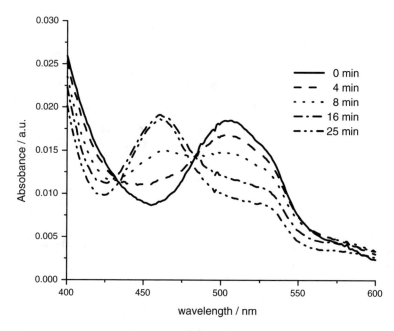

Fig. 5.9 Evolution of UV-Vis spectra of reduced technetium species in the presence of HNO_3 in 4 M H_2SO_4 (Chotkowski and Czerwiński 2016, reprinted with permission from Chotkowski and Czerwiński (2016) Copyright 2016 Creative Common License)

are generated. It should be stressed at this point that the literature descriptions of spectroscopic properties of ionic Tc(V) species given by various authors are inconsistent. Thus, e.g., Poineau et al. (2013) indicated that technetium(V) oxocations in sulfuric acid solutions generate a band at 695 nm.

Reactions of Tc–HNO_3–hydrazine system play a very significant role in the spent nuclear fuel reprocessing and thorough understanding of these processes is very important for the nuclear industry. The presence of hydrazine in the aqueous solutions prevents accumulation of HNO_2. The general equation illustrating this process can be written as follows (Kemp et al. 1993) (5.11):

$$N_2H_5^+ + 2HNO_2 \leftrightarrows N_2O + N_2 + 3H_2O + H^+ \tag{5.11}$$

Unfortunately, it turned out that technetium at intermediate oxidation states plays also a very important role in decomposition of the hydrazine-HNO_3 system. Tc (IV–VII) efficiently catalyzes the process of the hydrazine oxidation by nitric acid (Garraway and Wilson 1984; Kemp et al. 1993; Ozawa et al. 2003; Marchenko et al. 2008). Figure 5.10 presents a scheme of possible interaction in the Tc–HNO_3–hydrazine system. The values of the rate constants of reactions, which take place in this system are collected in Table 5.4. It is worth noting that unstable Tc(V) and Tc(VI) forms play a very important role in the evolution of the nitrous compounds.

Fig. 5.10 Reaction steps correspond to those reported in Table 5.3 (reprint with permission from Kemp et al. (1993) copyright 1993 Royal Society of Chemistry)

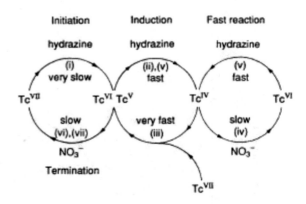

Table 5.4 Lists the values of the rate constants of each of its stages (Kemp et al. 1993)

Reaction	Best fit rate constant/dm^3 mol^{-1} s^{-1}
$Tc^{VII} + N_2H_5^+ \rightarrow$ $Tc^{VI} + N_2H_3^{\cdot} + H^+$	2.22×10^{-3}
$Tc^{V} + N_2H_5^+ \rightarrow$ $Tc^{IV} + N_2H_3^{\cdot} + H^+$	0.25
$Tc^{IV} + Tc^{VII} \rightarrow Tc^{VI} + Tc^{V}$	3.056
$Tc^{IV} + NO_3^- \rightarrow Tc^{VI} + NO_2^-$	1.25×10^{-3}
$Tc^{VI} + N_2H_5^+ \rightarrow$ $Tc^{IV} + N_2H_2 + 3H^+$	0.236
$Tc^{VI} + NO_3^- (+H^+) \rightarrow$ $Tc^{VII} + NO_2^-$	$5.55 \times 10^{-4} + K*$
$Tc^{V} + NO_3^- \rightarrow Tc^{VII} + HNO_2$	1.389×10^{-3}

$K*$ an acid concentration dependent rate constant (e.g., for 0.1 M HNO$_3$ $K = 30.1$; for 3 M HNO$_3$ $K = 1.43$)

Additional information on electrocatalytic properties of the reduced Tc species were delivered by Pertetrukhin et al. (2008). The experiments were carried out in HNO$_3$ + KNO$_3$ environment and in buffer solutions containing acetic and formic acids. The results enabled analysis of the reduction of Tc(VII) to Tc(III). On the basis of SDME results, the authors derived the following equation describing pH influence on $E_{1/2}$ (vs. Ag, AgCl) of this redox couple (5.12):

$$E_{1/2} = -0.0735 \cdot \text{pH} + 0.0696 \tag{5.12}$$

A significant increase in cathodic currents at potentials below 0.7 V (vs. Ag, AgCl) was attributed by these authors to reduction of HNO$_3$ to HNO$_2$. Such a behavior is probably related to catalytic properties of electrogenerated Tc(III) ions. These ions are rapidly oxidized by HNO_2 to Tc(IV) and Tc(V). The latter are reduced back to Tc(III) at the electrode surface. Tc(IV) and Tc(V) undergo a hydrolysis process, which leads to the formation of poorly soluble hydroxocomplexes. The current efficiency of the Tc electrodeposition on a graphite electrode ($S/V = 2500$ m^{-1}) in 1 M HNO_3 is low and for 2 h of the electrolysis it is equal to 9.05, 23.5 and 35.0% for − 0.5, 0.8 and −1.2 V, respectively.

Hoshi et al. (2007) investigated the electroreduction of the pertechnetates in 0.1– 9 M HNO_3 with addition of 0.1 M N_2H_4. The electroreduction was carried out at − 0.3 V (vs. Ag, AgCl) using a glassy carbon fiber electrode in a flow-type electrolysis cell. The solution remaining after completing the process was analyzed using UV–Vis, which detected $Tc(SCN)_6^{2-}$ complex. These authors identified Tc(IV) as the product of the pertechnetates electrochemical reduction. The more recent experiments performed in 6 M HNO_3 have shown that the reduction of the pertechnetates in solutions containing Pu and Np ionic forms leads to formation of Tc(IV) (Hoshi et al. 2007).

The electrochemical characterization of Tc–U–hydrazine systems in the basic environment was of interest to Peretrukhin et al. (1998). The authors applied polarographic measurements with a hanging mercury drop electrode. They observed a 3-electron irreversible reduction of Tc(VII) to Tc(IV). The half-wave potential of this redox couple was found to be equal to −0.69 V (vs. Ag, AgCl). A one-electron reduction of U(VI) to U(V) was observed at −0.87 V. An analysis of the difference in the potentials of reactions of both elements ($\Delta E =$ about 0.18 V) led these authors to the conclusion that Tc should interact with U(VI) and this process may partially proceed with participation of unstable U(V) intermediates. Very complex chemical processes were observed in 0.5–4 M NaOH solutions containing pertechnetates (1 × 10^{-5} ÷ 2 × 10^{-4} mol dm^{-3}), $N_2H_5NO_3$ (0.01–0.3 M) and sodium uranate. At room temperature, $N_2H_5NO_3$ completely reduced pertechnetates in such solutions within 5–10 min. An interaction of hydrazine with of 2 × 10^{-4} M Tc(VII) and 2 × 10^{-4} M U(VI) in 0.5 and 2 M NaOH led to formation of Tc(IV) hydrous oxide, which changes the solution color to brown-reddish. The complete reduction of the pertechnetates was confirmed by electrochemical measurements. A decrease in the concentration of U(VI) in the solution was attributed to its sorption and/or polymerization at the surface of $TcO_2 \cdot xH_2O$ as well as its partial reduction by hydrazine in the presence of Tc(IV) to $U_3O_8 \cdot nH_2O$ and $UO_2 \cdot nH_2O$.

The electrochemical studies on the pertechnetates reduction in an alkaline environment with high ionic strength solutions (up to 2 M NaOH and 5 M $NaNO_3$) have been reported by Chatterjee et al. (2018). The authors noted unexpected stability of Tc(VI or V?) forms. Also Chotkowski (2018) examined the mechanism of the pertechnetates reduction in NaOH solutions with concentration of up to 10 M NaOH. The results of both works are discussed in detail in Chap. 3.

The catalytic properties of technetium in the solutions containing hydrazine and selected actinides were also discussed by other authors (e.g. Koltunov et al. 1986; Kemp et al. 1993; German et al. 2011; Marchenko et al. 2008; Zhou et al. 2014). These works analyzed reduced technetium species generated under various experimental conditions and, as an effect, results reported by different groups differ from each other. For example, German et al. (2011) investigated the catalytic properties of technetium in Tc–(Np, Th, Zr)–$N_2H_5NO_3$–HNO_3. These authors obtained reduced Tc as a result of a reaction of the pertechnetates with an excess of hydrazine in 1.5 M HNO_3. In the next step, Np(V) was added to such prepared solution. The authors concluded that reactions in this system lead to the formation of Tc(V) and Np(IV). Reduction of Np(V) is catalyzed by Tc(IV) and the kinetics of this process follows Eq. (5.13):

$$-\frac{d[Np(V)]}{dt} = k^0[Tc(IV)][HNO_3]^2 \qquad (5.13)$$

where k^0 equals $14.8(2) \times 10^{-2}$ dm^6 mol^{-2} min^{-1}.

Also Zhou et al. (2014) investigated the reduction of Np(V) ions in the presence of Tc in HNO_3 and hydrazine containing solutions (5.14). The authors derived the following kinetic Eq. (5.14):

$$-\frac{d[Np(V)]}{dt} = k[Np(V)][Tc(IV)]^{0.8}[HNO_3]^{1.2} \qquad (5.14)$$

where k is equal to 28.5 ± 0.9 dm^6 mol^{-2} min^{-1} at 25 °C and the activation energy E_a equals 70.0 kJ mol^{-1}. These researchers pointed out that this process is accompanied by formation of Tc(V) and Tc(VI) intermediate species, which are involved in the oxidation of Np(IV) to Np(V). They also stated that this process is possible only when the standard redox potential of Tc(VI)/Tc(IV) system is sufficiently high, with the value of 1.291 V proposed by the authors.

Chotkowski (2018) investigated the electrochemical properties of technetium in the presence of neptunium(III, IV and VI) in 4 M H_2SO_4. A comparison of cyclic voltammetric curves recorded in acidic solutions of pertechnetates in the presence and absence of Np ions is presented in Fig. 5.11. NpO_2^+ ions are assumed to be the most stable forms of the neptunium present in slightly acidic solutions where their disproportionation, if possible, is considered as a slow process (Lemire et al. 2001). In strongly acidic media, on the other hand, Np^{4+} ions are the most stable neptunium species. The results obtained showed that the presence of Np^{4+} ions does not affect both the electroreduction of the pertechnetates and the oxidation of reduced forms of Tc.

The reduction of Tc(VII) to Tc(IV) forms is manifested by a cathodic peak at 0.5 ÷ 0.6 V (Fig. 5.11) and can be presented by reactions (5.15, 5.16):

$$TcO_4^- + 6H^+ + 2e^- \leftrightarrows TcO^{3+} + 3H_2O \qquad (5.15)$$

$$3TcO^{3+} + 3H_2O \leftrightarrows 2TcO^{2+} + TcO_4^- + 6H^+ \qquad (5.16)$$

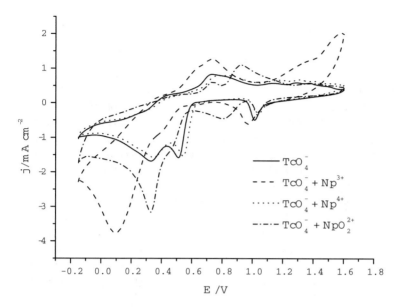

Fig. 5.11 Cyclic voltammograms of 5 mM TcO$_4^-$ and 11 mM Np(III, IV or VI) in 4 M H$_2$SO$_4$. Au electrode, room temperature, v = 200 mV s^{-1}, E versus Ag, AgCl$_{(KClsat.)}$ (reprinted with permission from Chotkowski (2018) copyright 2018 Elsevier)

A second broad and poorly shaped cathodic wave appears at potential range of 0 ÷ 0.45 V. Various electrochemical and chemical reactions of technetium compounds, mainly of those with oxidation states +IV and +III, may occur here. This includes electroreduction of Tc(IV) to unstable Tc(III) and possibly synproportionation of Tc(III) with pertechnetates(VII). One of the possible reactions can be schematically presented as (5.17, 5.18):

$$Tc(IV) + e^- \leftrightarrows Tc(III) \tag{5.17}$$

$$x Tc(III) + y Tc(VII) + z H^+ \leftrightarrows [Tc(III/IV)_{(x+y)}O_q)_{poly} + w H_2O \tag{5.18}$$

The presence of Np(VI) and Np(III) in the electrolyte solution has a significant impact on electrochemical reactions of Tc. The Np^{3+} ions exhibit strong reducing properties toward Tc(VII) and reduce them to Tc(IV) with a polymeric rather than a simple oxide structure. This reaction is manifested by appearance of a broad cathodic wave at potentials lower than 0.6 V (Fig. 5.11, dash line). A reduction of Tc(VII) starts immediately after mixing of Np^{3+} and TcO$_4^-$ solutions. This process is accompanied by a change in the colour of the solution. An excess concentration of NpO$_2^{2+}$ in respect to TcO$_4^-$ has also a strong impact on recorded CVs. A significant decrease in the intensity of the first TcO$_4^-$ reduction peak intensity due to the NpO$_2^{2+}$ presence is noticeable. The Tc(V) species generated at the electrode surface during this process are immediatelly oxidized by Np(VI).

Additional data related to the interactions between Np and Tc species were delivered by spectrophotometric experiments. Analyzed solutions contained NpO_2^{2+} ions and selected electrogenerated reduced Tc species, e.g., TcO^{2+}, Tc(III) and Tc(IV)-polymer. Oxidation of the Tc species leads to the formation of unstable Tc(V) and Np(V) ions. Tc(V) was characterized by a wave with a maximum at approx. 460 nm while the Np(V) ions generated a signal at ca. 980 nm. It has been found that the rate of the oxidation of the reduced Tc species strongly depends on their structure. The slowest process was observed for the polymeric forms of Tc(IV) while Tc(III) were oxidized by Np(VI) most rapidly.

In Japan, Masaki Ozawa and coworkers develops a concept of reprocessing of the spent nuclear fuel with the aim of recovering of elements important to the nuclear industry, e.g., rhenium or noble metals. This approach involves application of electrochemical techniques to deposit elements, such as Ru, Rh, Mo or Re, on cathodes directly from HLLW. The approach is referred to as Advanced-ORIENT Advanced Optimization by Recycling Instructive Elements Cycle strategy. Figure 5.12 presents a general scheme of this process.

Highly acidified waste containing FP is subjected to an electrolytic extraction. The rare metals that can be deposited on a Pt–Ti cathode in this process include noble metals (e.g., Ag, Rh, Pd), Se, Te and Tc. A research conducted by Ozawa et al. (2002, 2003, 2005, 2008) showed that the efficiency of their electrodeposition increases when Pd^{2+} is added to the system as a mediator (Fig. 5.13). Initially, the procedure involved a stepwise increase in the reduction currents according to the following scheme: 2.5 mA cm^{-2} (1 h) → 25 mA cm^{-2} (2 h) → 50 mA cm^{-2} (2 h) → 100 mA cm^{-2} (2 h). As a result, Ru–Rh–Pd–Re deposit was produced. Re was deposited as $ReO_{3(2)}$, Ru in its metallic form and Tc as TcO_2 (Ozawa et al. 2005). The maximum separation ratios of the metals of interest were as follows: >99% for Pd and Rh, 60% for Ru, 55% for Re and 25% for Tc (Ozawa et al. 2008). In the course of further research, the current densities flow sheet was modified to 2.5 mA cm^{-2} (1 h) → 75 mA cm^{-2} (2 h) → 100 mA cm^{-2} (4 h)) (Koyama et al. 2011). To increase the recovery ratio of $RuNO^{3+}$, TcO_4^- and Re O_4^- a decrease in the nitric acid concentration is required. The addition of Pd^{2+} to the galvanic bath (Pd:Tc = 5) increases the technetium electrodeposition efficiency and reduces the negative effect of HNO_3 content in the solution.

Ozawa and coworkers reported effective Tc deposition on platinum at potentials of hydrogen evolution region. They came to the conclusion that adsorbed hydrogen probably participates in the TcO_4^- ion reduction process, according to the Eq. (5.19):

$$TcO_4^- + 3H_{ads} + H^+ \leftrightarrows TcO_2 + 2H_2O \qquad (5.19)$$

They also observed that hydrogen evolution overpotential of Tc covered Pt electrode in 1 M NaOH is lowered by about 50 mV as compared with a bare platinum. Studies conducted in the 0.5 M HNO_3 environment (Koyama et al. 2011) have shown the hydrogen evolution overpotential on Tc–Rh is lower by about 25 mV

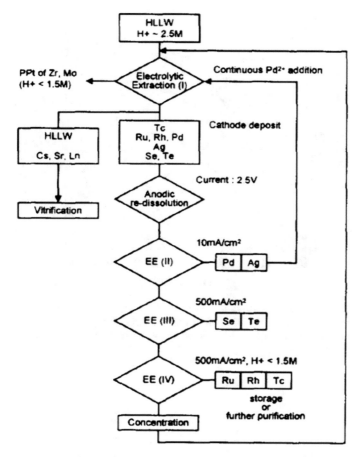

Fig. 5.12 Separation flowchart for fission-product rare elements by catalytic electrolytic extraction from HLLW (reprinted with permission from Ozawa et al. (2002) copyright 2002 Elsevier)

as compared with a pure Pt electrode. The authors suggested that the technetium catalytic activity toward hydrogen evolution is very high and can be even higher than that of rhenium. This points out to a possible application of Tc in hydrogen generating systems (Koyama et al. 2011).

Asakura et al. (2005) investigated the electrochemical reduction of the pertechnetates in nitric acid solutions as well as stability of Tc–Pd–Ru–Rh and Pd–Ru–Rh alloys. They found that addition of Tc to the alloy facilitates palladium dissolution. Cyclic voltammetry curves of the Pd–Ru–Rh alloy reveal two peaks: an anodic one at 0.89 V and the cathodic one at 0.5 V. For the Tc–Pd–Ru–Rh alloy, the potentials of both signals were shifted toward lower values by 40 and 20 mV for the cathodic and the anodic peak, respectively. Currents of these peaks decrease with the progress in continuous potential cycling of the Tc–Pd–Ru–Rh alloy (Fig. 5.14). A potential higher than 1.3 V (vs. Ag, AgCl, 3 M KCl) has to be applied in order to dissolve the

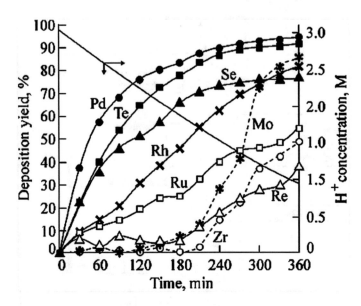

Fig. 5.13 Time dependency of deposition Yields on fission-product rare elements by electrolytic extraction in simulated HLLW (*C.D.*: 500 mA cm^{-2}, during 6 h electrolysis, *S/V*: 1/15 cm^{-1}, addition: continuously 2.53 g Pd^{2+} h^{-1}) (reprinted with permission from Ozawa et al. (2003) copyright 2003 Springer)

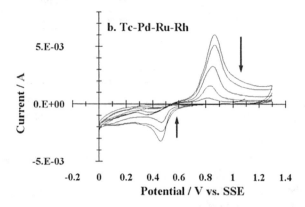

Fig. 5.14 Cyclic voltammograms of the deposits under multiple potential scans up to the tenth cycle. Anodic current positive. Initial potential: 1.1 V versus SSE. Initial scan direction: cathodic. Scan rate: 0.1 V s^{-1}. Potential range: 0–1.3 V versus SSE. Deposit from Tc–Pd–Ru–Rh solution. Electrolyte: 5.4 mM of Tc, 7.4 mM of Pd, 5.4 mM of RuNO^{3+} and 4.7 mM of Rh^{3+} in 3 M HNO$_3$ aqueous solution. Quantity of electricity for deposition: 378 mC. The first, the second, the fourth, the sixth and the tenth cycles (reprinted with permission from Asakura et al. (2005) copyright 2005 The Japan Society of Nuclear and Radiochemical Sciences)

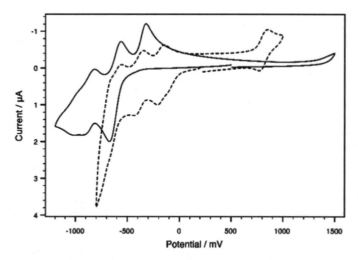

Fig. 5.15 Cyclic voltammograms of α1 (solid line) and $Tc^V O$-α1 (dashed line) in 0.1 M CH_3COONa/CH_3COOH containing 0.5 M Na_2SO_4, pH = 5.00. Working electrode, glassy carbon; auxiliary electrode, platinum wire; reference electrode, Ag/AgCl. Scan rate = 10 mV s^{-1} (reprinted with permission from McGregor et al. (2011) copyright 2011 American Chemical Society)

Tc-containing alloy. This value was 300 mV lower than that at which the effective dissolution of the Pd–Ru–Rh alloy was observed.

The reduction of the pertechnetates in nitric acid solutions carried out by Asakura et al. (2005) at the potential of −0.3 V led to the formation of a band at 482 nm. In contrast to Rotmanov et al. (2015) and Chotkowski and Czerwiński (2016) these authors attributed this signal to Tc(IV) complexes. In turn, the electrolysis of HNO_3 solution containing hydrazine and pertechnetates resulted in generation of a band at 430 nm, which was described as originated probably from Tc(III).

An interesting approach to immobilization of the technetium in a polyoxometalates matrix (POM) of a $[P_2W_{17}O_{61}]^{n-}$ type was presented by Burton-Pye et al. (2011). These authors analyzed several methods of preparation of such type of materials, including photochemical and electrochemical ones. The first one utilizes a photolytic reduction of the $-[P_2W_{17}O_{61}]^{10-}$ matrix to $-[P_2W_{17}O_{61}]^{12-}$ under UV irradiation conditions. In the next step the pertechnetates are reduced to Tc(V) in a two electron reaction with participation of the $-[P_2W_{17}O_{61}]^{12-}$ matrix. The second method is based on a electrochemical reduction of $-[P_2W_{17}O_{61}]^{10-}$ matrix at a glassy carbon electrode under applied potential of −220 mV (vs. Ag, AgCl reference electrode). Other papers devoted to this topic (McGregor et al. 2011, 2012) described the electrochemical properties of technetium in polyoxometalates Wells-Dawson isomers with a general formula of $K_{7-n}H_n[Tc^V O(\alpha1(or2)-P_2W_{17}O_{61})]$. CV curves typical for these compounds are presented in Fig. 5.15. The half-wave potential of the TcVI/TcV couple depends on the matrix used and varies from about 0.82 V to 1 V (vs. Ag, AgCl (3 M KCl)) for α2-$P_2W_{17}O_{61}$ and α1- $P_2W_{17}O_{61}$, respectively. These values were almost pH-independent for a broad pH values range (from 0 to

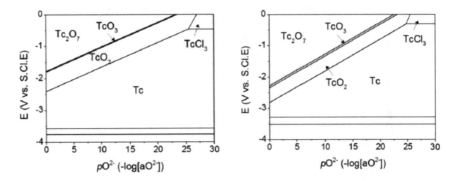

Fig. 5.16 Predominance diagrams of Tc in LiCl–KCl at 500 °C. and **b** in NaCl-KCl at 750 °C (reprinted with permission from Abdulaziz et al. (2016) Copyright 2016 Creative Common Licence)

7). On the other hand, the peak half-wave potential determined for the TcV/TcIV couple in the $\alpha 1$-$[P_2W_{17}O_{61}]^{10-}$ matrix decreases with the decrease in the acidity from 0.1 V for pH of 0 to -0.03 V for pH equal to 5.

A promising strategy of separation of technetium from other fission products utilizes an electrodeposition from nonaqueous solutions, e.g., molten salts. This process may include electrochemical dissolution of the SNF as one of the first steps. Although a fraction of the SNF containing noble metals and technetium precipitates as an insoluble waste, this strategy is still considered promising (Westphal et al. 2015). A further development of this method is complicated by the fact that only a handful of scientific papers deal with electrochemical behavior of technetium in molten salts. One of them is the work of Abdulaziz (2016), Abdulaziz et al. (2016), which presents the results of modeling the technetium behavior in nonaqueous eutectic systems (Fig. 5.16).

This author postulated the existence of technetium trioxide stability region in the phase diagrams. The experimental verification of this region is, however, very difficult due to a very narrow potential range when the stability is expected to be observed for a given pO^{2-} value. In contrast to the uranium or plutonium, the Tc phase diagrams do not include TcCl$_x$O$_y$ stability regions. The metallic technetium is formed in molten LiCl–KCl or NaCl–KCl salts in a wide range of O^{2-} concentration and at potentials which are higher than for the actinides. This observation is important from the point of view of a potential use of electrochemical separation of these elements. Abdulaziz (2016) derived the following equations which show a relation between the electrode potential and pO^{2-} for Tc compounds under equilibrium conditions (5.20–5.24):

$$TcO_2 + 4e^- \leftrightarrows Tc + 2O^{2-} \quad E = \frac{-\Delta G^{\ominus}}{4F} + \frac{2RT\ln 10}{4F} \cdot pO^{2-} \quad (5.20)$$

$$TcO_3 + 4e^- \leftrightarrows TcO_2 + 2O^{2-} \quad E = \frac{-\Delta G^{\ominus}}{2F} + \frac{RT\ln 10}{2F} \cdot pO^{2-} \quad (5.21)$$

$$Tc_2O_7 + 2e^- \leftrightarrows 2TcO_3 + O^{2-} \quad E = \frac{-\Delta G^\ominus}{2F} + \frac{RT\ln 10}{2F} \cdot pO^{2-} \quad (5.22)$$

$$Tc^{3+} + 3e^- \leftrightarrows Tc \quad E = \frac{-\Delta G^\ominus}{3F} \quad (5.23)$$

$$TcO_2 + 3Cl^- + e^- \leftrightarrows 2TcCl_3 + 2O^{2-} \quad E = \frac{-\Delta G^\ominus}{F} + \frac{RT\ln 10}{F} \cdot pO^{2-} \quad (5.24)$$

It is likely that the diagrams presented by Abdulaziz do not include all Tc forms that may occur in alkaline molten salts.

Volkovich et al. (2010) conducted research to characterize the electrochemical dissolution of metallic technetium in NaCl–2CsCl system at a temperature of 550 °C. The corrosion potential of the technetium film at this temperature was found to be equal to $0.50 \div 0.57$ V (vs. Ag, AgCl). The initial steps of this process are accompanied by dissolution of the metallic technetium and this process turns the color of the solution to yellow–brown and leads to formation of a vis shoulder at 420 nm. The UV–Vis spectrum of the solution recorded after completing the electrodeposition reveals bands at 300, 360 and 700 nm. The authors proposed that the overall process can be described by the general Eq. (5.25):

$$Tc \leftrightarrows Tc^{n+} + ne^- \quad (5.25)$$

The number of the electrons involved in oxidation of the metallic Tc (n) at the current density of 48 mA cm^{-2} was found to be equal to 3.0. The reduction of the current density to 18 mA cm^{-2} resulted in an increase in n to the value of 3.62. The zero-current potential for $TcCl_y^{n-}$ complexes in 16 mM Tc+NaCl–2CsCl solution was equal to 0.38 V (vs. Ag, AgCl or ~−0.9 V vs. Cl$^-$/Cl$_2$).

Figure 5.17 shows typical CVs recorded for metallic Tc in a molten salt. Two cathodic waves at potentials of ca. 0.5 V (c$_1$) and −0.4 V (c$_2$) are observed. Two corresponding anodic waves at ca. −0.3 V (a$_2$) and 0.6 V (a$_1$) are also recorded. Volkovich and coworkers pointed out that the anodic oxidation of Tc^0 leads to formation of products, which contain the technetium with two different oxidation states: +IV ($Tc_2Cl_8^{2-}$) and +III ($TcCl_6^{3-}$). It should be stressed, however, that formation of also $Tc_2Cl_8^{3-}$ containing Tc with mixed oxidation states cannot be ruled out. The redox system (a$_1$)–(c$_1$) is attributed to the reaction (5.26):

$$Tc^{4+} + ne^- \leftrightarrows Tc^{(4-n)+} \quad (5.26)$$

where (a$_2$)–(c$_2$) redox couple represents deposition and dissolution of the metallic technetium.

Volkovich drew attention to instability of ionic forms of technetium with lower oxidation states in the presence of oxygen.

Although the scientific literature widely reports information on the transport properties of numerous ionic species in molten salts (e.g., Okada 2002) there is a lack

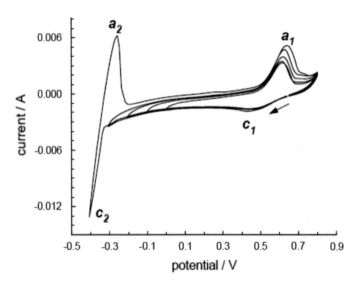

Fig. 5.17 Cyclic voltammograms of technetium in a NaCl-2CsCl melt at 550 °C recorded using a glassy carbon working electrode. The cathodic limit was progressively varied from 0 to −0.4 V (vs. Ag, AgCl). Scan rate 100 mV s^{-1}. Tc concentration ca. 0.026 mol dm^{-3} (reprinted with permission from Volkovich et al. (2010) copyright 2010 Creative Common Licence)

of reliable and comprehensive data concerning the conductivity and the diffusion coefficients of ionic forms of technetium in such systems. There are few publications, however, which deal with studies on transport properties of rhenium in molten salt systems. Rudenko et al. (2019), for example, studied the conductivity of molten eutectic CsCl–NaCl–KCl system containing Re(IV) with concentrations from 0 to 7.61 mol%. The authors pointed out that in such systems Re exists most likely in the form of $ReCl_6^{2-}$. They reported a linear relation between conductivity of these mixtures and the temperature, according to Eq. (5.27) where κ—molar conductivity (S cm^{-1}), T—temperature (K), a, b—empirical coefficients.

$$\kappa = (a \cdot T - b) \tag{5.27}$$

The a is equal to $(2.134 \pm 0.004) \times 10^{-3}$ while b depends on the Re concentration in the molten salt. The calculated b values were equal to 0.70374, 0.71118, 0.77743 and 0.96045 for the ReCl$_4$ content in the mixture of 1.55, 3.21, 5.32 and 7.61 mol%, respectively. It is expected that the behavior of Tc Cl_6^{2-} in this type of system should be very similar.

Salakhova (2014), in turn, summarized and reviewed publications devoted to the electrodeposition of rhenium and its alloys. One of the articles cited by this review reports values of the diffusion coefficient of Re(IV) ions in NaCl–KCl–ReCl$_4$ systems which are equal to 2.8×10^{-5} and 3.5×10^{-5} cm^2 s^{-1} for 790 and 840 °C, respectively. One may expect that Tc(IV) in these systems should exhibit similar values of the diffusion coefficient.

Although ^{99}Tc is a long-lived radioisotope, presence of high amounts of other short lived FP may result in a radiolysis of the materials containing the spent nuclear fuel. Therefore, an analysis of the electrochemical properties of technetium in nuclear fuel reprocessing systems must include evaluation of influence of the radiolysis on the stability of ionic forms of technetium. A detailed discussion of the radiolytic processes that involve technetium species goes beyond the subject of this monograph. Here, we focus only on selected general information on radiolytic stability of technetium(VII and IV) that are crucial for better understanding its redox chemistry.

A recent work of Ghalei et al. (2018) deals with characterization of radiolytic decomposition of $(NH_4)_2TcCl_6$ or TcO_4^- in bicarbonate and carbonate solutions. Under gamma irradiation the hexachlorotechnetates(IV) transform into carbonate complexes of Tc(IV) with $[Tc_2(\mu-O)_2(CO_3)_4(H_2O)_4]^{4-}$ structure proposed by the authors. The pertechnetates, on the other hand, are much more resistive to the radiolysis than the hexachlorotechnetates(IV) when the carbonates are present in the solution.

Earlier works (e.g. Lukens et al. 2001, 2002) indicated that irradiation of the pertechnetates in alkaline solutions in the presence of, e.g., nitrates leads to the formation of Tc(IV). Also Sekine et al. (2004) reported generation of TcO_2 colloids in solutions with pH > 2 and containing the pertechnetates which were subjected to irradiation with bremsstrahlung. Technetium with a low oxidation state, most likely Tc(I) in fac–$[Tc(CO)_3(gluconate)]^{2-}$ species, has been found in nuclear waste stored at Hanford Reservation (Lukens et al. 2004). In turn, Denden et al. (Denden et al. 2013) observed that under alpha irradiation technetium, initially present in highly acidic solutions as pertechnetates, transforms into oxopolymeric Tc(IV) species. More information on the radiolysis of various technetium species and actinides in alkaline solutions with various compositions can be found in WHC-EP-0901 report (Pikaev et al. 1996).

References

Abdulaziz R, Brown LD, Inman D et al (2016) Predominance diagrams of spent nuclear fuel materials in LiCl–KCl and NaCl–KCl molten salt eutectics. Int J Electrochem Sci 11:10417–10435

Abdulaziz R (2016) Electrochemical reduction of metal oxides in molten salts for nuclear reprocessing. PhD thesis, University College London, p 143

Adachi T, Ohnuki M, Yoshida N et al (1990) Dissolution study of spent PWR fuel: Dissolution behavior and chemical properties of insoluble residues. J Nucl Mater 174:60–71

Aihara H, Arai Y, Shibata et al (2016) Characterisation og the insoluble sludge from the dissolution of irradiated fast breeder reactor fuel. Proc Chem 21:279–284

Asakura T, Kim S-Y, Morita Y et al (2005) Study on electrolytic reduction of pertechnetate in nitric acid solution for electrolytic extraction of rare metals for future reprocessing. J Nucl Radiochem Sci 6(3):267–269

Bebko J (2011) Spectroelectrochemical investigations of pertechnetates reduction in the environment of sulfuric and nitric acid. Master thesis, University of Warsaw (in polish)

Box W (1968) Electrolyte for the electrodeposition of technetium. US Patent 3 374 157

Boyd GE (1959) Technetium and promethium. J Chem Edu 36(1):3–14

Boyd GE, Larson QV, Motta EE (1960) J Am Chem Soc 82:809–815

Bramman JI, Sharpe RM, Thom D, Yates G (1968) Metallic fission-product inclusions in irradiated oxide fuels. J Nucl Mater 25:201–215

Brewer L, Lamoreaux RH (1980) Thermochemical properties, in molybdenum: physico-chemical properties of its compounds and alloys. At Energy Rev, Spec Issue 7:1–191

Burton-Pye BP, Radivojevic I, McGregor D, Francesconi LC et al (2011), photoreduction of 99Tc pertechnetate by nanometer-sized metal oxides: new strategies for formation and sequestration of low-valent technetium. J Am Chem Soc 133:18802

Cartledge GH (1955) The pertechnetate ion as an inhibitor of corrosion. J Am Chem Soc 77:2658–2659

Cartledge GH (1971) The electrochemical behavior of technetium and iron containing technetium. J Electrochem Soc 118(11):1752–1757

Cartledge GH, Sympson RF (1957) The existence of a flade potential on iron inhibited by ions of the XO_4^- type. J Phys Chem 61(7):973–980

Chatterjee S, Hall GB, Johnsonet I et al (2018) Surprising formation of quasi-stable Tc(VI) in high ionic strength alkaline media. Inorg Chem Front 5:2081–2091

Chotkowski M, Czerwiński A (2012) Electrochemical and spectroelectrochemical studies of pertechnetate electroreduction in acidic media. Electrochim Acta 76:165–173

Chotkowski M, Czerwiński A (2016) Stability of technetium in the moderate oxidation states in acidic media. Annales UMCS Sectio AA LXXI(1): 141–149

Chotkowski M (2018) Redox interactions of technetium with neptunium in acid solutions. J Radioanal Nucl Chem 317:527–533

Compton RG, Sanders GHW (1996) Electrode potentials. Oxford Scientific Publications

Darby JB Jr, Norton LJ, Downey JW (1963) A survey of the binary systems of technetium with group VIII transition elements. J Less-Common Met 5:397–402

Denden I, Essehli R, Fattahi M (2013) Spectrophotometric study of the behaviour of pertechnetate in trifluoromethanesulfonic acid: effect of alpha irradiation on the stability of Tc(VII). J Radioanal Nucl Chem 296:149–155

de Zoubov, N, Pourbaix M (1966) Technetium Chapter IV, Section 11.2. In: Pourbaix M Atlas of electrochemical equilibria in aqueous solutions, Pergamon Press, p 298

Engelmann MD, Metz LA, Delmore JE (2008) Electrodeposition of technetium on platinum for thermal ionization mass spectrometry (TIMS). J Radioanal Nucl Chem 276(2):493–498

Ferrier M, Poineau F, Jarvinen GD et al (2013) Chemical and electrochemical behavior of metallic technetium in acidic media. J Radioanal Nucl Chem 298:1809–1817

Garcia-Garcia R, Ortega-Zarzosa G, Rincón ME et al (2014) The hydrogen evolution reaction on rhenium metallic electrodes: a selected review and new experimental evidence. Electrocatal 6(3):263–273

Garcia-Garcia R, Rivera JG, Antaño-Lopez R (2016) Impedance spectra of the cathodic hydrogen evolution reaction on polycrystalline rhenium. Int J Hydrogen Energy 41:4660–4669

Garraway J, Wilson PD (1984) The technetium-catalysed oxidation of hydrazine by nitric acid. J Less-Common Met 97:191–203

German KE, Obruchnikova YA, Tumanova DN et al (2011) Technetium catalytic effect and speciation in nitric acid solutions in presence of Np(V), Th(IV), Zr(IV) and reducing nitrogen derivatives. In: German KE, Myasoedov BF, Kodina GE, Maruk AY, Troshkina ID (eds) Book of proceedings, 7th ISTR July 4-8, 2011. Russia, Moscow, pp 114–120

Ghalei M, Vandenborre J, Poineau F (2018) Speciation of technetium in carbonate media under helium ions and γ radiation. Radiochim Acta 1–9. https://doi.org/10.1515/ract-2018-2939

Hoshi H, Wei Y-Z, Kumagai M (2007) Study on valence of Pu, Np and Tc in nitric acid after electrolytic reduction. J Alloys Compd 444–445:663–667

Inazawa S, Nitta K, Okada K et al (2016) molten salt bath, deposit, and method of producing metal deposit. US Patent 9 512 530 B2

Jaksic MM (2000) Volcano plots along the periodic table, their causes and consequences on electrocatalysis for hydrogen electrode reactions. J New Mat Electrochem Sys 3:167–182

Kemp TJ, Thyera SAM, Wilsonb PD (1993) The Role of Intermediate Oxidation States of Technetium in Catalysis of the Oxidation of Hydrazine by oxo-anions. Part I Nitrate ions. J Chem Soc Dalton Trans 2601–2605

Kleykamp H (1985a) The chemical state of the fission products in oxide fuels. J Nucl Mater 131:221–246

Kleykamp H (1985b) Composition and structure of fission products precipitates in irradiated oxide fuels: Correlaction with phase studies in the Mo–Ru–Rh–Pd and BaO–UO_2–ZrO_2–MoO_2 systems. J Nucl Mater 130:426–433

Kolman DG, Moore DP, Jarvinen GD et al (2012) the aqueous corrosion behavior of technetium–iron alloy materials. LA-UR-12–25629. https://doi.org/10.2172/1053896

Koltunov VS, Marchenko VI, Nikiforov AS (1986) The role taken by technetium in the oxidation-reduction processes used in irradiated-fuel technology. At Energy 43–51

Koyama S, Suzuki T, Mimura H et al (2011) Current status and future plans of Advanced ORIENT Cycle strategy. Prog Nucl Energy 53:980–987

Kuznetsov VV, Volkova MA, German KE et al (2020) Electroreduction of pertechnetate ions in concentrated acetate solutions. J Electroanal Chem 869:114090. https://doi.org/10.1016/j.jelechem.2020.114090. Accessed 20 June 2020

Láng GG, Horányi G (2003) Some interesting aspects of the catalytic and electrocatalytic reduction of perchlorate ions. J Electroanal Chem 552:197–211

Lemire RJ, Fuger J, Nitsche H et al (2001) Chemical thermodynamics of neptunium and plutonium, vol 4. Elsevier, Amsterdam, pp 91–104

Lukens WW, Bucher JJ, Edelstein NM et al (2001) Radiolysis of TcO_4^- in alkaline, nitrate solutions: reduction by NO_3^{2-}. J Phys Chem A 105:9611–9615

Lukens WW, Bucher JJ, Edelstein NM et al (2002) Products of pertechnetate radiolysis in highly alkaline solution: structure of $TcO_2 \cdot xH_2O$. Environ Sci Technol 36:1124–1129

Lukens W, Shuh D, Schroeder N et al (2004) Identification of the non-pertechnetate species in hanford waste tanks, Tc(I)-carbonyl complexes. Los Alamos Technical Report 38:229–233

Magee RJ, Cardwell TJ (1974) Rhenium and Technetium in: Bard AJ (ed) Encyclopedia of electrochemistry of the elements. vol II, Marcel Dekker, pp 126–189

Marchenko VI, Zhuravleva GI, Dvoeglazov KN et al (2008) Behaviors of plutonium and neptunium in nitric acid solutions containing hydrazine and technetium ions. Theor Fund Chem Eng 42(5):733–739

Masahira Y, Ohishi Y, Kurosaki K et al (2015) Effect of Mo content on thermal and mechanical properties of Mo–Ru–Rh–Pd alloys. J Nucl Mater 456:369–372

Maslennikov A, Masson M, Peretroukhine V et al (1998) Technetium electrodeposition from aqueous formate solutions: electrolysis kinetics and material balance study. Radiochim Acta 83:31–37

Maslennikov A (2012) Electrochemistry of actinides and selected fission products in the head end of spent nuclear fuel reprocessing. Procedia Chem 7:39–44

Mausolf E, Poineau F, Hartmann T et al (2011) Characterization of electrodeposited technetium on gold foil. J Electrochem Soc 158(3):E32–E35

McGregor D, Burton-Pye BP, Howell RC et al (2011) Synthesis, structure elucidation, and redox properties of 99Tc complexes of lacunary wells-dawson polyoxometalates: insights into molecular 99Tc-metal oxide interactions. Inorg Chem 50:1670–1681

McGregor D, Burton-Pye BP, Mbomekalle IM et al (2012) 99Tc and Re incorporated into metal oxide polyoxometalates: oxidation state stability elucidated by electrochemistry and theory. Inorg Chem 51:9017–9028

OECD (2012) NEA/NSC/WPFC/DOC(2012)15 Spent nuclear fuel reprocessing flowsheet. Nuclear Energy Agency. p 10

Okada I (2002) Transport properties of molten salts. In: Bockris JO'M, Conway BE, White RE (eds) Modern aspects of electrochemistry, vol 34, Kluwer, pp 119–204

Ozawa M, Shinoda Y, Sano Y (2002) The separation of fission products rare elements toward bridging the nuclear and soft energy systems. Prog Nucl Energy 40(3–4):527–538

Ozawa M, Ishida M, Sano Y (2003) Strategic separation of technetium and rare metal fission-products in spent nuclear fuel e solvent extraction behavior and partitioning by catalytic electrolytic extraction. Radiochem (Radiokhim) 45(3):225–232

Ozawa M, Suzuki T, Koyama S et al (2005) Separation of rare metal fission products in radiactive wastes in new direction of their utilization. Prog Nucl Energy 47(1–4):462–471

Ozawa M, Suzuki T, Koyama S (2008) new back-end cycle strategy for enhancing separation, transmutation and utilization of materials (Adv.-ORIENT cycle). Prog Nucl Energy 50:476–482

Peretrukhin VF, Silin VI, Kareta AV et al (1998) Purification of alkaline solutions and wastes from actinides and technetium by coprecipitation with some carriers using the method of appearing reagents: Final Report Final report of Institute of Physical Chemistry of Russian Academy of Sciences Contract with DOE, PNNL 1998

Perterukhin VF, Moisy F, Maslennikov AG (2008) Physicochemical behavior of uranium and technetium in some new stages of the nuclear fuel cycle. Russ J Gen Chem 78(5):1031–1046

Pikaev AK, Gogolev AV, Kryutchkov SV et al (1996) Radiolysis of Actinides and Technetium in Alkaline Media. WHC-EP-0901, Westinghouse Hanford Corporation

Poineau F, Weck PF, Burton-Pye BP et al (2013) Reactivity of $HTcO_4$ with methanol in sulfuric acid: Tc-sulfate complexes revealed by XAFS spectroscopy and first principles calculations. Dalton Trans 42:4348–4352

Poineau F, Gray K, Ebert W (2014) Electrochemical corrosion studies for modeling metallic waste form release rates. Project No. 12–4026. Final report. Nuclear Energy University Program USA

Poineau F, Koury DJ, Bertoia J et al (2016) Electrochemical studies of technetium-ruthenium alloys in HNO3: implications for the behavior of technetium waste forms[1]. Radiochemistry 59(1):41–47

Rard JA, Rand MH, Anderegg G et al (1999) Chemical thermodynamics of technetium, vol 3. Elsevier

Rotmanov KV, Maslennikov AG, Zakharova LV, Goncharenko YuD, Pertetrukhin VF (2015) Anodic dissolution oof Tc metal in HNO3 solutions. Radiochemistry 57(1):26–30

Rudenko A, Isakov A, Apisarov A et al (2019) Liquidus temperature and electrical conductivity of molten eutectic CsCl–NaCl–KCl containing $ReCl_4$. J Chem Eng Data 64:567–573

Salakhova E (2014) The electrochemical deposition of rhenium chalcogenides from different electrolytes. J Chem Eng Chem Res 1(3):185–198

Sekine T, Narushima H, Suzuki et al (2004) Technetium(IV) oxide colloids produced by radiolytic reactions in aqueous pertechnetate solution. Colloids Surf. A Physicochem Eng Asp 249:105–109

Skriver HL, Rosengaard NM (1992) Surface energy and work function of elemental metals. Phys Rev 46(11):7157–7168

Sympson RF, Cartledge GH (1956) The mechanism of the inhibition of corrosion by the pertechnetate ion. IV. Comparison with other XO_4^{n-} inhibitors. J Phys Chem 60(8):1037–1043

Szabó S, Bakos I (2000) Electroreduction of rhenium from sulfuric acid solutions of perrhenic acid. J Electroanal Chem 492(2):103–111

Taylor CD (2011) Surface segregation and adsorption effects of iron–technetium alloys from first-principles. J Nucl Mater 408:183–187

Trasatti S (1972a) Discussion of the electrochemical behavior of technetium and iron containing technetium. Cartledge GH (pp 1752–1758, vol 118, no 11)] Discussion Section. J Electrochem Soc 119(12):1696–1697

Trasatti S (1972b) Work function, electronegativity, and electrochemical behavior of metals. III Electrolytic hydrogen evolution in acid soultions. J Electroanal Chem Interfacial Electrochem 39:163–184

Trasatti S, Petri OA (1991) Real surface area measurements in electrochemistry. Pure Appl Chem 63(5):711–734

Volkovich VA, Vasin BD, Griffiths TR (2010) Electrochemical and spectroscopic properties of technetium in fused alkali metal chlorides. ECS Trans 33(7):381–390

Voltz RE, Holt ML (1967) Electrodeposition of Tc^{99} from aqueous solution. J Electrochem Soc 114(2):128–131

Westphal BR, Frank BR, McCartin WM et al (2015) Characterization of irradiated metal waste from the pyrometallurgical treatment of used EBR-II fuel. Metall Mater Trans A 46A:83–92

Wotteen CB (1977) Method for prevention of fouling by marine growth and corrosion utilizing technetium-99. US Patent 4 017 370

Yamanaka S, Kurosaki K (2003) Thermophysical properties of Mo–Ru–Rh–Pd alloys. J Alloys Compd 353:269273

Zakharov EN, Bagaev SP, Kudryavtsev VN et al (1991) On the possibility of a hydride phase formation under Tc-99 electrodeposition. Zashch Met 27(6):1024–1026

Zhou X, Ye G, Zhang H et al (2014) Chemical behavior of neptunium in the presence of technetium in nitric acid media. Radiochim Acta 102(1–2):111–116

Zhuz Q, Wang S-Q (2016) Trends and regularities for halogen adsorption on various metal surfaces. J Electrochem Soc 163(9):H796–H808

Zilberman BY, Pokhitonova YA, Kirshin MY et al (2007) Prospects for development of a process for recovering technetium from spent fuel of nuclear power plants. Radiochem 49(2):156–161

Chapter 6
Determination of Trace Amounts of Tc by Electrochemical Methods

Today, technetium is continuously released into the natural environment as a result of human activity in the fields of the nuclear industry and the nuclear medicine. It is estimated the uranium isotopes enrichment processes carried out in 1980s led to the release of the technetium in the amount which is an equivalent to 8 GBq of ^{99}Tc per 1000 MWe of power produced in a nuclear reactor using such enriched fuel (Desmet and Myttenaere 1986). Additional about 20 GBq of Tc has been released into the environment as a result of diagnostic examinations conducted with application of ^{99m}Tc labeled radiopharmaceuticals (Shi et al. 2012). Results of determination of the ^{99}Tc concentration in water, soil and living organisms clearly indicate that this element is present in measurable quantities in the contemporary environment. As an example, the technetium content in seawater and soil in Japan was determined at a level of 1/10 μBq dm^{-3} and 4/88 mBq kg^{-1}, respectively (Takahashi 2017). For this reason, there is a growing interest in development of fast and low-cost methods of determination of this element in environmental samples. Very popular are methods based on application of mass spectroscopy (MS) as has been presented in a review article by Shi et al. (2012).

Alongside frequently used MS-based approaches, e.g., ICP–MS (Inductively coupled plasma mass spectrometry) also radiometric techniques, such as LSC (Liquid scintillation counting) or NAA (Neutron activation analysis), are widely applied to the Tc quantification. In general, these techniques exhibit a low detection limit of Tc, usually at a level of $10^{-12}/10^{-15}$ g. Advanced mass spectrometry (RIMS: Resonance ionization mass spectrometry) is able to determine Tc at a level of 10^7 atoms in a sample (Rimke et al. 1990). Application of RIMS, however, may be expensive and this is the main disadvantage of the method. It is worth to mention that methods based on a gravimetric analysis of practically insoluble high molecular mass salts of technetium, such as tetraphenylarsonium pertechnetate, which determine Tc at a level of mg (Jasim et al. 1960) are no longer developed and seem to be of only historical interest today.

© Springer Nature Switzerland AG 2021
M. Chotkowski and A. Czerwiński, *Electrochemistry of Technetium*,
Monographs in Electrochemistry, https://doi.org/10.1007/978-3-030-62863-5_6

Electroanalytical methods create interesting alternative to expensive advanced mass spectrometry methods. They allow to determine the technetium at a μmol dm^{-3} and even nmol dm^{-3} level.

6.1   Acidic and Neutral Solutions

Lewis et al. (1985) determined electrochemically technetium concentration in acetate buffer solutions with pH of 4.8. Anodic stripping voltammperometry (ASV) and chronocoulometry on a wax-impregnated graphite electrode (WIGE) were selected as the electroanalytical techniques. The preconcentration of technetium at the electrode was carried out at -0.90 V versus Hg, Hg$_2$Cl$_{2(\text{NaClsat.})}$ for 10 min under stirring conditions. Application of more negative potentials was excluded due to interferences with currents from the hydrogen evolution. The researchers assumed that the electrode reaction associated with the reduction of Tc(VII) leads mainly to formation of TcO$_2$. The ASV curves recorded in an acetate buffer reveal existence of two current waves, denoted as peaks A and B in Fig. 6.1. In a solution containing 0.08 M NaCl, the first of these peaks is observed at 0.13 V and the second one at 0.33 V.

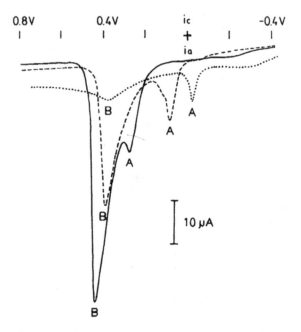

Fig. 6.1 Effect of chloride concentration on anodic stripping voltammograms with chloride concentrations of: (—) 0.0 M, (---) 0.5 M, ($\cdots$) 5.0 M 1.9 μM KTcO$_4$ in 0.5 M, pH 4.8 acetate buffer, WIGE, scan rate 50 mV s^{-1}, deposition potential -0.90 V; deposition time 10 min (reprinted with permission from Lewis et al. (1985). Copyright 1985 Elsevier)

An increase in the Cl⁻ ions concentration in the solution decreases intensity of the peak B although its potential remains approximately unchanged. In contrast to the peak B, the potential of the peak A is shifted toward lower values when the NaCl concentration increases while its intensity remains relatively unchanged. Such influence of the chlorides on the electrochemical behavior of the technetium species can be understood when the discussed process leads to formation of technetium-chloro or technetium-oxochloro species rather than pertechnetates. Lewis et al. determined technetium in a buffer solution without chlorides by means of chronocoulometric measurements. They reported the detection limit equal to $(1.00 \pm 0.01) \cdot 10^{-7}$ mol dm^{-3}. The procedure required deposition of Tc at -0.90 V for 10 min followed by shifting the potential to -0.10 V and finally to 0.60 V. The measurements were carried out at -0.10 V rather than at -0.90 V in order to reduce contributions from the double layer charging on the blank chronocoulogram. Lewis et al. concluded that the proposed method of the Tc determination is affected by several factors, including the pH value and the presence of the chlorides which influence height of the current peaks. Changes in the electrode surface due to aging processes, on the other hand, affect reproducibility of the results.

A combination of a high pressure liquid chromatography (HPLC) with electrochemical detection of technetium was applied by Lewis et al. (1983) for analysis of solutions eluted from a ^{99}Mo/^{99m}Tc generator. The surface of the stationary phase of the HPLC contained $-$NH$_2$ functional groups. The tests were carried out in acetates solutions or in phosphate buffers (pH $= 5/6$) using carbon electrodes or static mercury drop electrode (SMDE). The Tc detection limits obtained for the carbon and mercury electrodes were equal to $8.5 \cdot 10^{-9}$ and $2.1 \cdot 10^{-8}$ mol dm^{-3}, respectively. It is worth mentioning that the technetium solubility in the liquid mercury is low. Kozin estimated its value as equal to $1.1 \cdot 10^{-9}$ at% at 25 °C (Guminski and Galus 1986). Therefore, application of the Hg as a liquid electrode is intended to obtain a reproducible surface rather than Hg(Tc) amalgam.

Using adsorption stripping voltammetry at HMDE Friedrich and Ruf (1986) determined the technetium concentration at the level of approx. 10^{-11} g ml^{-1} ($\sim 10^{-10}$ mol dm^{-3}). The system was preconcentrated for 120 s at the potential of -0.4 V versus Ag, AgCl. The measurements were performed out using the differential cyclic voltammetry technique with a modulation amplitude of -40 mV, an electrode polarization rate of 15 mV s^{-1} and a potential ranging from -0.8 V to -1.5 V. The influence of the pulse amplitude (from -40 to -300 mV) on the high of the stripping voltammetric peak of technetium turned out to be relatively small. The measurements were carried out in solutions containing H$_2$SO$_4$ with a concentration of about 5 mM and various concentrations of SCN⁻. An analysis of the currents due to technetium reaction at -1.27 V revealed that the highest signal is observed for the thiocyanates concentration of 0.66 mM (Fig. 6.2).

Friedrich and Ruf stated that application of solutions with high acid concentration is disadvantageous due to a possible reaction between the Tc(VII) ions with the metallic mercury electrode. Therefore, the experiments were carried out in solutions with a pH above 2. The presence of NaCl and Na$_2$SO$_4$ significantly influenced the height of the recorded current signals. Thus, when the concentration of these salts

Fig. 6.2 Influence of thiocyanate concentration on the height of the stripping voltammetric peak of technetium. [Tc] = 1.28 · 10^{-8} mol dm^{-3}; deposition time = 120 s (reprinted with permission from Friedrich and Ruf (1986). Copyright 1986 Elsevier)

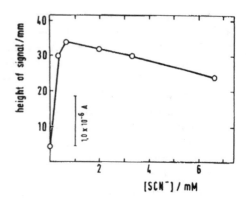

Fig. 6.2 Influence of thiocyanate concentration on the height of the stripping voltammetric peak of technetium. [Tc] = 1.28 · 10^{-8} mol dm^{-3}; deposition time = 120 s (reprinted with permission from Friedrich and Ruf (1986). Copyright 1986 Elsevier)

is as high as 0.06 mol dm^{-3} the currents due to the Tc reduction are smaller than those recorded in a salt-free solution by ca. 55% and 80% for the chlorides and the sulphates, respectively.

A modified tri-n-octylphosphinooxide (TOPO) glassy carbon electrode (GCE) was used by Torres Llosa et al. (1988a, b) for determination of Tc in 3 M HCl solutions. Stable HTcO$_4$·2TOPO complex was the technetium species analyzed by the researchers. Application of stripping voltammetry allows to determine the technetium concentration at a level of 10^{-8} mol dm^{-3}. The analysis focused on the height of the Tc(VII) reduction peak at the potential of -0.35 V versus Ag, AgCl. The best results were obtained using a high pulse amplitude (-100 mV or higher) and a scan rate of 60 mV s^{-1}. According to the researchers, Tc(IV) can be complexed by TOPO. Unfortunately, potentials of reduction of both discussed technetium forms, i.e., Tc(VII) and Tc(IV), are very close to each other making its separation very difficult. The effect of interferences from U(VI) was also discussed (Figs. 6.3 and 6.4). The uranium content at a level similar to or higher than that of the technetium precludes the determination of Tc by this method. Torres Llosa et al. stated that also other metals that are extracted by the TOPO from strongly acidic solutions, e.g., Fe, Mo, Zn, Zr, Nb and Sn, may affect the process of the technetium determination.

Using stripping chronopotentiometry at a glassy carbon electrode in 5 mM H$_2$SO$_4$, Ruf (1988) determined the technetium concentration at a level of approx. 10^{-7} mol dm^{-3}. Preconcentration of Tc at the electrode was performed at a potential of -0.7 V versus Ag, AgCl for 5 min. Ruf also analyzed the influence of oxidants, such as Ce^{4+} or H$_2$O$_2$, on the quantitative determination of Tc in the solution. Thus, for 10 mM Ce(IV) the transition time varies linearly with the Tc concentration until the latter reaches approx. 1 · 10^{-6} mol dm^{-3}. This researcher stated that the transition time of the oxidation of the Tc deposit obtained in the preconcentration step depends on whether the solution subjected to the preconcentration procedure contained Tc(VII) or Tc(IV). They also noted that the perrhenates did not affect the Tc determination because the signal associated with oxidation of the Re deposit generated during the preconcentration step was observed at ca. 0.37 V, which is a significantly lower value than the potential of the Tc oxidation. Ruf reported that in contrast to the behavior

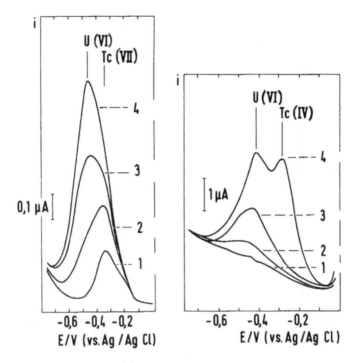

Fig. 6.3 (left panel) Stripping voltammogram for $3.6 \cdot 10^{-8}$ M Tc(VII) solution with different contents of uranium(VI): (1) 0; (2) $4 \cdot 10^{-7}$ M; (3) $8 \cdot 10^{-7}$ M; (4) $1.2 \cdot 10^{-6}$ M. Enrichment time, 10 min; pulse amplitude, -100 mV, scan rate, 15 mV s^{-1}. (right panel) Stripping voltammogram for Tc(IV) and U(VI) solutions. Curves: (1) supporting electrolyte only; (2) $8 \cdot 10^{-8}$ M U(VI); (3) $2.4 \cdot 10^{-7}$ M U(VI); (4) $2.4 \cdot 10^{-7}$ M U(VI)/Tc(IV) (reprinted with permission from ref. Torres Llosa et al. (1988a). Copyright 1988 Elsevier)

observed for the Ru(III), the presence of Fe(III), Sb(III), Zr(IV), Mn(II), Zn(II) did not affect the technetium determination as long as their total concentration is lower than that of the Tc. Cu, Cd and Pb, on the other hand, are reported to interfere with the Tc measurements. The presence of sulfates and nitrates at a concentration of 1 mol dm^{-3} as well as presence of Mo(VI) and W(VI) ions leads to a reduction in the transition time of the Tc oxidation.

An interesting approach to determination of technetium presented by German et al. (2005). These authors examined TcO$_4^-$ ion-selective PVC membrane electrodes using quaternary alkylammonium or phosphonium bromides and pertechnetates as ionophores. The detection limit was determined to be $9 \cdot 10^{-7}$ mol dm^{-3}. The results showed that these types of the electrodes are stable over broad range of pH, from c.a. 1 up to 13. Moreover an 10^4 fold excess of nitrates, bromides and 10^6 sevenfold excess of sulfates, phosphates, chlorides ions did not interfere with the Tc determination.

In an extremely acidic environment of 12 M H$_2$SO$_4$, Chotkowski et al. (2018) determined the technetium concentration on the basis of the height of the current peak due to reduction of Tc(VII) to Tc(V). Derivative pulse voltammetry curves

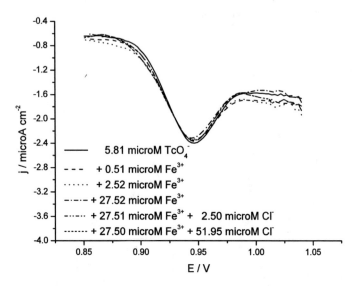

Fig. 6.4 Derivative pulse voltammetry (DPV) recorded on Au electrode in $12\,M\,H_2SO_4 + 5.81\,\mu M$ TcO_4^- ions in the absence and presence of various concentrations of Fe^{3+} and Cl^- ions. Pulse amplitude: 50 mV, scan rate: 5 mV s^{-1} (reprinted with permission from Chotkowski et al. (2018). Copyright 2018 Elsevier)

recorded by these authors revealed a well-developed signal in the potential range of 0.8/0.9 V versus Ag, $AgCl_{(KClsat.)}$. The detection limit was determined to be $9 \cdot 10^{-7}$ mol dm^{-3}. The addition of interfering agents, such as Fe^{3+} and Cl^- ions, at a μmol dm^{-3} level did not affect the results of the technetium determination in the sample (Fig. 6.4).

Application of the electrochemical methods of determination of the technetium concentration in the radiopharmaceuticals obtained from the physiological serum just before administration to the patient was reported by Herlem et al. (2015). The solution eluted from a $^{99}Mo/^{99m}Tc$ generator contained $^{99m}TcO_4^-$ ions, which were complexed with 3,3-diphosphono-1,2-propanedicarbonic acid (DPD, also known as Teceos). Application of the derivative pulse stripping voltammetry and platinum ultramicroelectrodes with a diameter of a 5 μm allows quantification of $NaTcO_4 \cdot 2DPD$ and $NaTcO_4$ with the detection limits at a level of 5 nmol dm^{-3} and 30 nmol dm^{-3}, respectively (Fig. 6.5b, c). The authors have found that the background signal recorded for the 5 μm electrode was much smaller than the one observed for the electrode diameter of 72 μm. The technetium preconcentration step was carried out at a potential of -1 V versus Ag, AgCl for 30 s, the applied pulse amplitude and the scan rate were equal to 25 mV and 25 mV s^{-1}, respectively. Re-dissolution of TcO_2 deposit in the solution containing only pertechnetates without complexing agents was observed at a potential of 0.54 V. The latter was shifted toward more positive value of 0.64 V in the presence of DPD. (Fig. 6.5a). Herlem et al. also stated that this methodology can be also applied to other radiopharmaceuticals (Myoview and Stamicis). In Myoview + $NaTcO_4$ solutions the electrochemical signals were

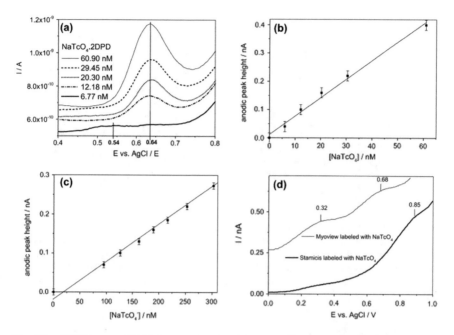

Fig. 6.5 **a** DPSV scans of NaTcO$_4$ complexed by DPD. **b** Calibration curve of NaTcO$_4$ · 2DPD. **c** Calibration curve of NaTcO$_4$ **d** DPSV scan of Myoview (15 nM) and Stamicis (20 nM) ligands labeled with NaTcO$_4$ (reprinted with permission from Herlem et al. (2015). Copyright 2015 Elsevier)

observed at 0.32 V and 0.68 V (Fig. 6.5d) while in the case of solutions containing Stamicis + NaTcO$_4$ only a single peak appeared at 0.85 V.

Chatterjee et al. (2011) constructed a spectroelectrochemical sensor for [Tc (dmpe)$_3$]$^{2+/+}$ (dmpe = 1,2-bis (dimethylphosphino) ethane) detection in 0.1 M KNO$_3$. This sensor consisted of an optically transparent ITO (indium-tin oxides) material which surface was covered with a thin (typically 315 nm) layer of a sulfonated polystyrene-block-poly (ethylene-ran-butyle) block-polystyrene (SSEBS) film. The cationic form of [Tc(dmpe)$_3$]$^+$ was found to be strongly bounded by the SSEBS. Moreover, the [Tc(dmpe)$_3$]$^{2+}$ form has been found to be highly emissive in aqueous solutions. An analysis of the intensity of Tc(II)-complex emission line (λ_{em} = 660 nm) allowed to quantify the technetium with the detection limit of 24 nmol dm^{-3}. The measurement procedure required preconcentration by ion-exchange for 30 min. This is done by immersing the ITO/SSEBS sensor in a solution containing [Tc(dmpe)$_3$]$^+$. After that, the system was washed first with water and then with 0.1 M KNO$_3$. The luminescence was excited with a laser with a wavelength of 532 nm. The [Tc(dmpe)$_3$]$^{2+}$ was generated during the electrode polarisation at 0.6 V versus Ag, AgCl$_{(3\ M\ NaCl)}$. The measurements of the emission at 660 nm were carried out immediately after termination of the electrode polarization at 0.6 V versus Ag, AgCl$_{(3\ M\ NaCl)}$ (Fig. 6.6).

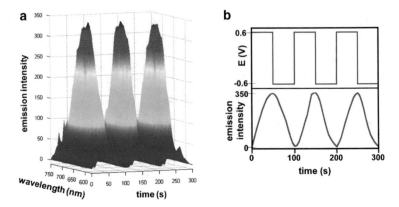

Fig. 6.6 **a** Optical modulation profile of emission intensity ($\lambda_{ex} = 532$ nm) upon reversible inter-conversion of $[Tc(dmpe)_3]^+$ (nonemissive) and $[Tc(dmpe)_3]^{2+}$ (emissive) on the application of varying step potential (shown in **b**) plotted versus wavelength and time; **b** (blue−) plot of potential applied during the optical modulation with respect to time; (red−) plot of modulation of emission intensity at 660 nm as a function of applied potential (reprinted with permission from Chatterjee et al. (2011). Copyright 2011 American Chemical Society)

6.2 Alkaline Solutions

The traces of the Tc in alkaline solutions were determined by Astheimer and Schwochau (1967). Using the hanging mercury dropping electrode, they precon-centated the Tc during electrolysis at a potential of −1.0 V versus Ag, AgCl. The technetium content in the sample was calculated on the basis of the first oxidation peak, which appeared at −0.33 V (Fig. 6.7). A linear relation between its height and the Tc concentration was obtained when the latter were in the range from 10^{-4} to 3

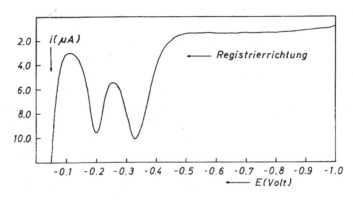

Fig. 6.7 Inverse polarogramm of $6.0 \cdot 10^{-5}$ M KTcO$_4$ in 1 M NaOH, temp. 25 °C, potential versus Ag, AgCl, diameter of Hg drop ~0.8 mm, enrichment time 4 min., scan rate 1 V min^{-1} (reprinted with permission from Astheimer and Schwochau (1967). Copyright 1967 Elsevier)

$\cdot 10^{-7}$ mol dm^{-3}. The second oxidation peak observed at -0.2 V allowed to determine the technetium at the concentration range from 10^{-4} to $4 \cdot 10^{-6}$ mol dm^{-3}. The authors stated that an 10^4 fold excess of ReO$_4^-$ or MoO$_4^{2-}$ ions did not interfere with the Tc determination when the content of the latter was equal to at least 0.5 μg.

A different approach to the analysis of the traces of the technetium was proposed by El-Reefy et al. (El-Reefy 1990). These researchers used the technetium extraction as one of the steps in the process of determination of this element. They utilized the fact that tetraphenylarsonium strongly extracts the TcO$_4^-$ from an alkaline or a slightly acidic aqueous phase to a chloform-based organic phase. The extraction process was carried out for 10 min. using the aqueous phase containing 0.01 M HCl and Tc(VII) while the organic phase was composed of 10 mM tetraphenylarsonium dissolved in chloroform. The volume ratio of the organic to the aqueous phase was equal to 1:3.

The preconcentration of the Tc on a hanging mercury drop electrode (HMDE) was carried out in an organic phase (6 cm^3), which was alkalized by addition of 3 cm^3 of 0.1 M NaOH. The addition of 9 cm^3 of ethanol to such obtained solution resulted in the formation of a homogeneous phase, which was used in the respective electrochemical measurements. The process of the Tc preconcentration on the HMDE electrode was carried out at a potential of -1.6 V versus Ag, AgCl for 2 min. Using a current wave observed at -0.26 V on differential pulse cathodic stripping voltammetry curves, El-Reefy et al. were able to quantify the Tc with the detection limit of $3 \cdot 10^{-8}$ mol dm^{-3} (Fig. 6.8). The best sensitivity was obtained using a negative pulse amplitude of -40 mV and a scan rate of 20 mV s^{-1}.

Fig. 6.8 Cathodic stripping voltammogram of different pertechnetate concentrations in TPAC — chloroform phase mixed with NaOH and ethanol, $T_{dep} = 2$ min; $U_{dep} = -1.6$ V versus Ag, AgCl, $U_{dep} = -4.0$ mV and scan rate $= 20$ mV s^{-1} (reprinted with permission from El-Reefy et al. (1990). Copyright 1990 Elsevier)

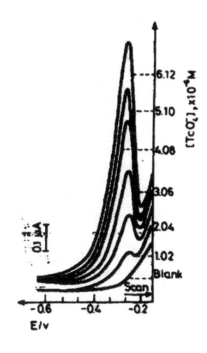

The concentration of the technetium in 0.1 M NaOH was analyzed by Abuzwida (Abuzwida 1996). This researcher analyzed anodic stripping voltammetry curves recorded at glassy carbon electrodes and found a linear dependence of the current at -0.25 V versus Ag, AgCl on the Tc concentration when the latter was in the range of $2 \cdot 10^{-6}/5 \cdot 10^{-8}$ mol dm^{-3}. This signal was attributed to the oxidation of Tc(IV) initially deposited on the electrode. Preconcentration was made at a potential of -1.4 V for a period from 30 s to 5 min. Abuzwida reported that U(VI) interferes with the Tc and, in order to reduce these interferences, the uranium concentration should not be higher than $1 \cdot 10^{-7}$ mol dm^{-3} for the examined range of the Tc concentrations.

References

Abuzwida MA (1996) Anodic stripping voltammetry of technetium alkaline media. In: Third Arab conference on the peaceful uses of atomic energy. Damascus 9–13 Dec 1996, AAEA. https://inis. iaea.org/collection/NCLCollectionStore/_Public/31/065/31065341.pdf. Accessed 15 May 2019

Astheimer L, Schwochau K (1967) Zur invers-polarographie des technetiums. J Electroanal Chem 14:240–241

Chatterjee S, Del Negro A, Edwards MK et al (2011) Luminescence-based spectroelectrochemical sensor for [Tc(dmpe)3]2+/+ (dmpe = 1,2-bis(dimethylphosphino)ethane) within a charge-selective polymer film. Anal Chem 83:1766–1772

Chotkowski M, Wrzosek B, Grden M (2018) Intermediate oxidation states of technetium in concentrated sulfuric acid solutions. J Electroanal Chem 814:83–90

Desmet K, Myttenaere K (1986) Technetium in the environment. Elsevier

El-Reefy SA, Ruf H, Schrob K (1990) Catodic stripping voltammetry in the tethaphenylarsonium chloride/chloroform extract at a hanging mercury drop electrode. J Radioanal Nucl Chem 141(1):179–183

Friedrich M, Ruf H (1986) Assay of extremely low technetium concentration by adsorption strpping voltammetry at the HMDE after reaction with thiocyanate. J Electroanal Chem 198:261–268

German KE, Dorokhov AV, Kopytin AV et al (2005) Quaternary alkylammonium and alkylphosphonium pertechnetates: application to pertechnetate ion-selective electrodes. J Nucl Radiochem Sci 6(3):217–219

Guminski C, Galus Z (1986) Radioactive elements. In: Hirayama C, Galus Z, Guminski C (eds) Metals in mercury, IUPAC solubility data series, vol 25. Pergamon Press, p 421

Herlem G, Angoue O, Gharbi T et al (2015) Electrochemistry of pertechnetate on ultramicroelectrode: a new quality control for radiopharmaceuticals manufactured at hospitals in nuclear medicine. Electrochem Comm 51:76–80

Jasim F, Magee RJ, Wilson CL (1960) Chemical analysis on the microgram scale VI. The ultramicrogravimetric determination of technetium and rhenium. Microchim Acta 5–6:721–728

Lewis JY, Zodda JP, Deutsch E et al (1983) Determination of pertechnetate by liquid chromatography with reductive electrochemical detection. Anal Chem 55:708–713

Lewis JY, Pinkerton TC, Deutsch E et al (1985) Stripping chronocoulometry for the determination of pertechnetate. Anal Chim Acta 167:335–342

Rimke H, Herrmann G, Mang M et al (1990) Principle and analytical applications of resonance ionization mass spectrometry. Microchim Acta III:223–230

Ruf H (1988) Quantitative analytical assay of small amounts of technetium by stripping chronopotentiometry at a glassy carbon electrode. J Electroanal Chem 241:125–131

Shi K, Hou X, Roos P et al (2012) Determination of technetium-99 in environmental samples: a review. Anal Chim Acta 709:1–20

Takahashi Y (2017) Technetium. In: White WM (ed) Encyclopedia of geochemistry. Springer, p 1422

Torres Llosa JM, Ruf H, Schorb K et al (1988a) Stripping voltammetric determination of traces of technetium with glassy carbon electrode coated with tri-n-octylphosphine oxide. Anal Chim Acta 211:317–323

Torres Llosa JM, Ruf H, Schorb K et al (1988b) Stripping voltammetry assay of trace technetium with a TOPO coated glassy carbon electrode. J Res Nation Bureau Stand 93(3):493–495

About the Authors

Maciej Chotkowski is Assistant Professor at the Faculty of Chemistry of the University of Warsaw (UW), Poland. Following chemistry studies at UW he worked for his Ph.D., which was granted in 2008. In 2019 he acquired the title of a habilitated doctor (*doctor habilitowany*). In 2004–2012 he worked in the Industrial Chemistry Research Institute, Poland. He is a specialist in application of electrochemical and spectroelectrochemical methods in investigations of the redox behavior of technetium, manganese and rhenium compounds. He has published research papers concerning mainly the fundamental properties of inorganic ionic species of technetium and manganese in aqueous media. He has participated in research projects in nuclear chemistry focused on industry and medicine and he was a member of steering or advisory committees of scientific conferences, e.g. of the *3rd Academic Symposium on Nuclear Fuel Cycle* (2015 Japan), and the *10th International Symposium on Technetium and Rhenium. Science and Utilization* (2018 Russia). He teaches a large variety of regular courses and laboratories at UW dealing with physical and nuclear chemistry. He is also very active in introducing innovative concepts of teaching and adaptation of chemical experiments to the needs of visually disabled students.

© Springer Nature Switzerland AG 2021
M. Chotkowski and A. Czerwiński, *Electrochemistry of Technetium*,
Monographs in Electrochemistry, https://doi.org/10.1007/978-3-030-62863-5

Andrzej Aleksander Czerwiński is a Professor at the University of Warsaw Faculty of Chemistry. Following his studies of chemistry at UW, he obtained his Ph.D. in 1974. In 1989 received his scientific degree, D.Sc. (habilitation). In 1978–79 and 1983–85, he worked with Harry B. Mark, Jr. (University of Cincinnati, USA) and in 1990–1991 with Roberto Marassi (University of Camerino, Italy). After graduation, he initially worked in the automotive industry, followed by the Nuclear Research Institute working on the production of closed radiation sources. Since 1971 he has been employed at the Department of Chemistry of the University of Warsaw. From 1999 through 2019 he worked in the Industry Chemistry Research Institute, where from 2007 to 2015 he was the Chairman of the Scientific Council. He has led more than 20 competitive research national and international projects, has supervised 22 Ph.D. theses, has registered 35 patents and has published more than 240 research works in indexed journals (Hirsh Index 37) as well as 16 textbooks from chemistry and energy for university and high school students. His research contributions have been to the fields of nuclear chemistry, electrochemistry and teaching chemistry.

In the course of these contributions, he has investigated the mechanism of catalytic oxidation processes of fuels (simple organic compounds) on metal electrodes of the platinum group where the tested compounds were labeled with radioactive carbon isotope C-14. Using radioactive isotopes, he has studied homogeneous electron exchange in V^{2+}/V^{3+} and Eu^{2+}/Eu^{3+} systems. Together with Dr. M. Chotkowski initiated spectro-electrochemical studies of rhenium and technetium in the laboratory and it is continued intensively by the team of Dr. M. Chotkowski.

A. Czerwiński is currently focusing on electrochemical power sources. His achievements include the development of a new carbon lead—acid battery with significantly increased capacity, the use of ionic liquids as an electrolyte in nickel-hydride batteries, which together with the modification of the anode surface of alloys lead to better power and corrosion resistance in these kind of cells. In 1978, he introduced Nafion into thin-layer electrodes. He developed a new LVE (limited volume electrode) method successfully used to study hydrogen sorption in materials like palladium, its alloys, and transition metal compositions used in the construction of nickel-metal hydride cells. He has published more than 120 papers on this subject. Many previously unknown in the literature properties of these systems have been discovered using this method. He is a laureate of many awards and distinctions for scientific achievements and inventions, including the Prime Minister award for outstanding scientific and technical achievement. He is currently the head of the Laboratory of Electrochemical Power Sources at the Department of Chemistry, University of Warsaw.

About the Series Editor

Fritz Scholz is Professor at the University of Greifswald, Germany. Following studies of chemistry at Humboldt University, Berlin, he obtained a Dr. rer. nat. and a Dr. sc. nat. (habilitation) from that University. In 1987 and 1989 he worked with Alan Bond in Australia. His main interest is in electrochemistry, electroanalysis and history of science. He has published more than 350 scientific papers and books. He is editor and co-author of the book "Electroanalytical Methods" (Springer, 2002, 2005, 2010, and Russian Edition: BINOM, 2006), coauthor of the book "Electrochemistry of Immobilized Particles and Droplets" (Springer 2005), co-editor and co-author of the "Electrochemical Dictionary" (Springer, 2008; 2nd ed. 2012), and co-editor of volumes 7a and 7b of the "Encyclopedia of Electrochemistry" (Wiley-VCH 2006) and other books. In 1997 he has founded the *Journal of Solid State Electrochemistry* (Springer) and serves as Editor-in-Chief since that time. In 2014 he has founded the journal *ChemTexts—The Textbook Journal of Chemistry* (Springer). He is editor of the series "Monographs in Electrochemistry" (Springer) in which modern topics of electrochemistry are presented. Scholz introduced the technique 'Voltammetry of Immobilized Microparticles' for studying the electrochemistry of solid compounds and materials, he introduced three-phase electrodes to determine the Gibbs energies of ion transfer between immiscible liquids, and currently he is studying the interaction of free oxygen radicals with metal surfaces, as well as the interaction of liposomes with the surface of mercury electrodes in order to assess membrane properties. Two books are devoted to the history of science: the autobiography of Wilhelm Ostwald (Springer 2017) and the title "Electrochemistry in a divided world" (Springer 2015).

© Springer Nature Switzerland AG 2021
M. Chotkowski and A. Czerwiński, *Electrochemistry of Technetium*,
Monographs in Electrochemistry, https://doi.org/10.1007/978-3-030-62863-5

Index

A

Abundance
  in earth's crusts, 3
  in oceans, 3
  in stellar, 2
Activation energy, 40, 41, 128
Activity, mean molar
  $HClO_4$, 20
  $HReO_4$, 20
  $HTcO_4$, 20, 32
  $NaTcO_4$, 32
Adsorption
  process, 51, 70, 113, 114
  rhenium compounds, 26, 40
  technetium compounds, 26
Alloys
  Mo–Ru–Rh–Pd, 111
  Ru–Rh–Pd–Re, 130
  Tc–Fe, 113, 123
  Tc–Pd–Ru–Rh, 131
  Tc–Ru, 121–123

B

Batteries, 16

C

Catalytic properties, 109, 121, 124, 126–128
Chronocoulometry, 75, 80, 144
Clusters, 48, 49, 60, 61
Conductance
  Tc complexes, 61–62, 77
  TcO_4^-, 32
Corrosion
  Fe, 113
  metallic Tc, 112, 121–123, 135
  Mo alloys, 111
  protection, 113
  Tc-Fe, 113, 123
  Tc-Ru, 122, 123

D

Differential Pulse Voltammetry (DPV), 148, 151
Diffusion coefficient
  manganese species, 17
  rhenium species, 17, 60, 80, 136
  TcO_4^-, 17, 33
  technetium species, 17, 61, 80, 136
Dimerization, 11, 43, 54
Discovery
  davyum, 1
  masurium, 1
  nipponium, 1
  panorama, 2
  technetium, 1–3
Disproportionation
  Np(V), 128
  Re(IV), 22
  Re(VI), 21
  Tc(V), 36, 81
  Tc(VI), 36, 52–55, 62

E

E_h-pH diagram, 12–15
Electrochemical Quartz Crystal Microbalance (EQCM), 22, 23, 26
Electrode material, influence, 23, 27, 39
Electronegativity, 11, 12, 61
E_L, parameters values
  for ligand classes, 103

© Springer Nature Switzerland AG 2021
M. Chotkowski and A. Czerwiński, *Electrochemistry of Technetium*,
Monographs in Electrochemistry, https://doi.org/10.1007/978-3-030-62863-5

for ligands, 96–102

F
Fission products, 5, 109, 110, 132, 134

G
Gamma rays
    irradiation, 137
    technetium isotopes, 4

H
Hydrazine, 109, 125–128, 133
Hydrogen evolution, 12, 15, 21, 22, 24, 113,
        117, 118, 130, 131, 144

I
Inductively coupled plasma mass spec-
        troscopy (ICP-MS), 143
Interferences
    antimony, 147
    cadmium, 147
    cerium, 146
    chlorides, 144–148
    copper, 147
    hydrogen peroxide, 146
    iron, 148
    lead, 147
    manganese, 147
    molybdenum, 151
    nitrates, 147
    rhenium, 151
    ruthenium, 147
    uranium, 146, 147, 152
    zinc, 146, 147
    zirconium, 146, 147
Ion selective membrane, 147
Isotopes
    instability, 3, 5
    production, 4, 5
    properties, 4

K
Kinetic parameters, 35, 36, 55, 58, 63, 128

L
Ligands, organic
    (acac)$_2$en, 83, 93
    bpm, 85, 93

bpy, 73, 75, 84, 89, 93
(brac)$_2$en, 83, 93
(buac)$_2$en, 83, 93
(bzac)$_2$en, 83, 93
CH$_3$CN, 77, 90, 93
CF$_3$COO, 72, 84, 93
CN-C$_6$H$_{11}$, 71, 84, 93
CN-tert-butyl, 71, 84, 93
DBAP, 92, 93
DBCat, 76, 91, 93
dedc, 87, 93
depe, 20, 70, 72, 80, 82, 83, 87, 93
diars, 20, 70, 82, 93
diphos, 82, 93
dioxime:, 75
dmdc, 87, 93
DMP, 81
dmpe, 19, 20, 70, 72–74, 82, 83, 85–87,
        93
DMSA, 81
dppb, 70, 82, 86, 87, 93
dppbt(h), 74, 86, 93
dppe, 20, 70, 71, 82–84, 93, 96
en, 92, 93, 98
FcCS$_2$, 76, 91, 93
gluconate, 93, 137
HDPE, 79, 80
(H)DPhF, 77, 90, 93
morphbtu, 75, 88, 93
L^1, 91, 93
H$_2$L^{1b}, 88, 93
L$^{'}$, 91, 93
L$^{''}$, 86, 93
L^2, 76, 91, 93
L^3, 76, 91, 93
L$_B$^{1}, 76, 91, 93
L$_B$^{2}, 76, 91, 93
L$_B$^{3}, 76, 91, 93
L$_T$, 76, 91, 93
L$^{3'}$, 76, 91, 93
lut, 89, 93
L$_q$, 91, 93
Me$_2$bpy, 73, 75, 88, 93
noet:, 74
PEt$_2$Ph, 84, 93
PEt$_3$, 75, 83, 93
phen, 84, 89, 93
pic, 89, 93
PMe$_2$Ph, 70, 77, 84, 90, 93
PMe$_3$, 77, 90, 93
pmdc, 87, 93
PnAO, 84, 93
P(OEt)$_2$Ph, 78

PPh$_3$, 88, 93
PR$_2$S, 93
py, 75, 88, 89, 93
9S3, 66, 93
(sal)$_2$en, 83, 93
SC$_6$H$_4$-p-Cl, 74, 86, 93
SC$_6$H$_4$-p-OCH$_3$, 74, 86, 87, 93
SC$_6$H$_4$-p-t-Bu, 86, 93
SC$_6$H$_5$, 74, 86, 93
SCH$_2$C$_6$H$_4$-p-OCH$_3$, 74, 86, 93
SCH$_2$C$_6$H$_5$, 74, 86, 93
SCP, 87, 93
(S$_2$CPh)(S$_3$CPh)$_2$, 91, 93
SES, 87, 93
S-n-C$_3$H$_7$, 86, 93
t-butyl$_3$terpy, 88, 93
TCTA, 77, 90, 93
tdt, 86, 93
terpy, 84, 88, 89, 93
tmeda, 75, 88, 93
tmf$_2$enH$_2$, 75, 93
tmpp, 75, 88, 93
tn, 81, 92, 93
TpXPC, 88, 93
Ligands, inorganic
    Re(CO)$_5$X, 62
    TcBr$_6^{2-}$, 45–47
    TcCl$_6^{2-}$, 35, 45–47, 49, 50, 61, 62, 137
    TcI$_6^{2-}$, 60
    [Tc$_x$X$_y$]$^{n-}$, 48, 49
    TcNBr$_4^-$, 47
    TcNCl$_4^-$, 47
    [Tc(NCS)$_6$]$^{3-}$, 61, 62
    [Tc(NH$_3$)$_4$(NO)H$_2$O]$^{2+}$, 50
    [Tc(NH$_3$)$_4$(NO)F]$^+$
    [Tc(NO)(NCS)$_5$]$^{2+}$, 62
    TcOBr$_4^-$, 47
    TcOCl$_4^-$, 47, 50
    TcO(CN)$_5^{2-}$, 60
    Tc(OH)$_p$(CO$_3$)$_q^{4-p-2q}$, 59
    [Tc$_6$Q$_8$(CN)$_6$]$^{4-}$, 61–62
    [Tc$_6$S$_8$X$_6$]$^{4-}$, 61–62
Liquid chromatography, 145
Liquid scintillation counting (LSC), 143

M
Manganese
    electrochemistry, 11, 12, 16, 17
    ionic forms, 16, 17, 27
    oxides, 12, 16
Mercury electrode, 18, 21, 36, 52, 79–81,
        127, 145, 150, 151

Metallic technetium
    conductivity, 5
    dissolution, 112, 120–123, 135
    electrodeposition, 114–117, 119
    hydrogen evolution, 15, 113, 117, 118
    physicochemical properties, 5, 6
Microelectrodes, 71, 148
Molten salts
    conductance of Re(IV), 136
    diagrams of Tc, 134
Molybdenum
    electrochemistry, 11–13
    ionic forms, 11–13, 27
    isotopes, 3, 4

N
Neptunium
    interaction with technetium, 128–130
Neutron activation Analysis (NAA), 143
Nitric acids
    decomposition, 124, 125
    redox potentials, 124, 125
Nuclear fuel, 109–112, 130, 137

O
Oxides
    Tc$_2$O$_7$, 22, 135
    Tc$_2$O$_5$, 22
    TcO$_3$, 22, 36
    TcO$_2$, see Tc(IV)

P
Palladium, 110, 131
Pitzer parameters, 31–33
Plutonium, 5, 111, 112, 127, 134
Polymeric forms (or species), 11, 13, 14, 38,
        42–44, 63, 120, 124, 129, 130, 137
Polyoxometalates, 133
Protonation
    TcO$_4^-$, 53
    TcO$_4^{2-}$, 53

R
Resonance ionization mass spectrometry,
        143
Rhenium
    electrochemistry, 16, 17, 19–22, 24–26,
        39, 61, 62, 81, 115, 119, 130
    E$_h$-ph diagram, 14

ionic forms, 12, 16–18, 27, 136
nucleation mechanism, 22
oxides, 14, 22, 23, 26, 119
Ruthenium
    electrochemistry, 11, 12, 122
    E_h-ph diagram, 14
    ionic forms, 11–13, 27
    isotopes, 3

S
Serum, 148
Solubility
    pertechnetates, 6, 7
    Tc in Hg, 145
Spectroelectrochemical sensor, 149
Stability, ionic forms
    oxochlorotechnetium(III/IV), 48
    TcO^{3+}, 15, 22, 36, 128
    TcO_4^{2-}, 52–55
    TcO^{2+}, 13, 15, 34, 43, 128
    Tc^{3+}, 13, 16, 41–43, 95
    Tc^{2+}, 12, 44, 95
Stripping chronopotentiometry, 146
Stripping voltammetry, 145, 146, 148, 151,
    152
Synproportionation, 43, 129

T
Tafel slope, 113, 117, 121
Tc(III)

oxidation, 16, 40, 41, 76, 95
polymerization, 11, 12, 43, 44
stability, 95
Tc(IV)
    equilibrium data, 34
    polymerization, 12, 13, 27, 37, 41–43,
        54, 62, 112
    stability, 13, 15, 48, 51, 112–114, 134
    standard reduction potentials of tech-
        netium coulpes, summarized, 63

U
Uranium
    fuel, 110
    interaction with technetium, 127
    isotopes, 143

W
Work function, 113, 114

Y
Yield
    fission products production, 5
    techentium isotopes production, 5

Z
Zirconium, 4, 109, 115, 128, 146, 147

Printed in the United States
by Baker & Taylor Publisher Services